DATE DUE

"I'm Mighty Apt to Make a Million or Two"

NELLIE CASHMAN

and the North American Mining Frontier

GREAT WEST AND INDIAN SERIES, VOLUME SIXTY-THREE

"I'm Mighty Apt to Make a Million or Two"

NELLIE CASHMAN

and the North American Mining Frontier

By

DON CHAPUT

WESTERNLORE PRESS . . . 1995 . . . TUCSON, ARIZONA

Library of Congress Catalog Number 95-60928
ISBN 0-87026-093-6

PRINTED IN THE UNITED STATES OF AMERICA BY WESTERNLORE PRESS

TABLE OF CONTENTS

		Page
Preface		vii
Acknowledgements		ix
I.	Getting Started	1
II.	Canadian Challenge	17
III.	Arizona Calls	27
IV.	Weak Legends, Great Adventures	55
V.	Nellie Buries Tombstone	75
VI.	From Desert Sand to Glacier Peaks	89
VII.	Fulfillment on the Koyukuk	121
VIII.	End of the Paystreak	141
Notes to the Chapters		157
Bibliography		175
Index		183

PREFACE

THE history of the American West has been my focus for some years, especially mining, exploration, and military affairs. Nellie Cashman is no departure from this Western concentration. Instead she is one of the ultimate stampeders.

Most authors who have considered Nellie's career have emphasized the women's angles, such as her help in establishing hospitals and churches and her caring for miners in trouble. The image of Nellie "with a cup of soup for a sick miner" may be accurate but woefully incomplete. Nellie was a person of the frontier, hardened by five decades of toil, speculation, troubles and triumphs in the roughest mining camps of North America. The fact that she was decent and caring should not obscure Nellie's lifelong goals: she was in the hills, mountains, deserts, and frozen ground in order to find precious metals, which she could then give away to those in need.

Gold rushes, stampedes, and boom towns attracted hundreds of different personality types, and hundreds of thousands of people. The amount of enthusiasm, optimism, and audacity shown by these stampeders has always fascinated me. Many were gamblers, men of the green cloth; others were lawyers and officers of the law; still others were dreamers, teachers, speculators, clergymen, merchants, women of easy virtue.

The Earp brothers rushed to Deadwood, Tombstone, Nome, and Goldfield. E. J. "Lucky" Baldwin, Tex Rickard, Dave Neagel, Rex Beacher, Jack London, and "Arizona Charlie" Meadows were in the forefront of Western mining rushes. These people and their kind were pikers, short-timers, compared to Nellie Cashman.

She was at home in the Nevada Desert, San Francisco, British Columbia, Tombstone, Baja California, New Mexico, the Klondike, and north of the Arctic Circle in Alaska. Consider what climates these various places bring to mind, and consider that Nellie's stampede days began in 1872 and only ended when she died in 1925.

As familiar as I am with the mining history of the Americas, I know of no other stampeder who showed such endurance, such optimism, and who earned such glowing praises from her fellow prospectors and miners than Nellie Cashman.

ACKNOWLEDGEMENTS

CARL Chafin provided essential guidance and information on this Nellie Cashman quest. He is well known as an authority on 19th Century Arizona. Not so well known is that Carl is also well grounded in Alaska's past. He has been my severest critic and my greatest contributor.

Sister Mary Loyola of the Sisters of St. Ann in Victoria, British Columbia, Canada, became fascinated with Nellie's career in the early 1970s, and she brought together important manuscripts and other records from the scattered missions of that order; in addition she obtained copies of primary materials from other libraries, archives, and government offices in the United States and Canada. I was privileged to use these records in 1994 with the help of the current archivist, Sister Margaret Cantwell, a veteran of four decades of Alaska service and author of *North to Share,* a history of that order's work in the Yukon and Alaska.

My wife Toni did her best to work with my prose, which tends to be not just informal but downright folksy. Her help with my research, computer problems, and search for illustrations has been immeasurable.

The Library staff of the Natural History Museum, Los Angeles, coordinated many of the unusual loans from other institutions as well as providing essential mining and political publications and maps of Western North America. In Bisbee, Arizona, Cochise County Recorder Christine Rhodes filled several important gaps in my Nellie research; Bisbee is the county seat, replacing Tombstone, where Nellie spent so many exciting years.

Nellie's Irish origins were revealed through the efforts of Tim Cadogan of the Cork County Library, who enlisted the support of Canon P. Troy, P.P., of Midleton, home of the Cashmans.

The only known photo of a youthful Nellie, was taken in San Francisco in 1874, just before she left for the British Columbia gold fields. (Archives, Sisters of St. Ann, Victoria)

I

GETTING STARTED

THE North American dreams of Nellie Cashman began in the southwest corner of Ireland, in County Cork. Yet, the details and the chronology of her origins have been obscured by time, by the devastating impact of the potato famine, and by Nellie's own contradictory comments over the years.

The name Cashman is geographically distinct, limited mostly to the Counties of Cork and Kerry. The correct Gaelic version of the name was O'Kissane, which gradually became simplified to Kissane, later Anglicized to Cashman. A usual translation is "one who seized people by the head," such as the local bailiff or the tax collector.[1]

In the many interviews and statements of and about Nellie over the years, there is not a clue about her father. Recently uncovered evidence from County Cork has provided many pieces of information about the Cashmans, although some vital links still remain obscure.

They came from the parish of Midleton, a small farming settlement on the coast, a few miles to the east of the City of Queenstown; in the 1840s Queenstown was known as Cove, was renamed Queenstown at the time of Queen Victoria's visit in 1849, and obtained its current name of Cobh after Irish Independence in the 1920s. Nellie frequently mentioned "Queenstown," when people asked for her place of birth. This made some sense, as that was the most well known city in County Cork, and was only a few miles from her birthplace at Midleton.

Patrick Cashman married Fanny Cronin in Midleton on October 9, 1842. Their ages are not given, but neither was likely very young. Fanny, in fact, may have been forty years old at this time. In the following year, September 9 of 1843, a son Patrick was born and baptized, but apparently died soon thereafter. Ellen Cashman, the Nellie of our interest, was baptized in Midleton on October 15, 1845.

From this point on the Midleton records, so well kept and so useful, are no longer helpful. Some facts mixed in with a few conjectures suggest the following for a reasonable chronology for the next few years. Another daughter, Frances, was born to the Cashmans within a few years, and she and her older sister Nellie would be practically inseparable for the next quarter-century. The Cashmans, though, seem to have left Midleton by this time, and before the end of the 1850s, there is no longer a Patrick Cashman. Fanny Cashman had become a widow, and whether farm accident, disease, or some tragedy related to the potato famine took her husband has not come down to us.

From the evidence in the parish records, and from the few comments from Nellie over the years, it seems that around 1850 the widow Fanny Cashman with young daughters Nellie and Fanny decided to put the famine and hopelessness of Ireland behind them, and they headed for the New World.[2]

Such a departure would have placed the Cashmans in the crush of emigration following those disastrous potato famine years of the late 1840s, when hundreds of thousands of Irishmen headed for England, the United States, Canada, and Australia. Of the southern counties, Cork was the most severely impacted by the famine and for thirty years was the leading emigrant county. For example, in the early 1850s, 90,500 fled Cork, while the second leading county in the south, Tipperary, had an emigrant total of 60,000.[3]

The reasons for the famine were many, and these have led to

an extensive literature. The unexpected potato disease, the English restructuring of the industrial-trade network, and the English insistence for years of ignoring or denying the Irish tragedy led to disaster in Ireland. One of the more crushing blows to Ireland was the increased availability of factory-made goods, mostly from England. Thus the village craftsmen were thrown double-trouble: imported goods meant their services were not needed, and the potato disease, which eliminated tens of thousands of small farmers, meant that their meager skills would no longer be needed in the agricultural community.[4]

The greatest distress came to those least able to handle it, the impoverished small farmers, or those urban dwellers with few skills little needed in the rapidly changing industrial environment. The Cashmans, we can safely conclude, from the hardest hit County of Cork, were among the classes hardest hit who decided that the future was elsewhere, not in Ireland.

Patrick Cashman was dead, although a few chroniclers who have examined Nellie's life suggest that perhaps the family left Ireland as a unit, and he died at sea. If so, the information was never imparted to Nellie, as she frequently stated that "my mother, who was a widow, brought me here to America."[5] So, the widowed Frances Cashman left Ireland with her two baby girls, sometime around 1850, for the United States.

To some this may seem incredulous. If we assume that the Cashmans were impoverished, and throw in widowhood and two baby girls for complications, we might have difficulty in writing the emigration-to-the-New-World scenario. Yet, among the Irish, this was practically the norm.

Female emigration was higher from Ireland than from any other European country in the nineteenth century. As a comparison, fifty-two percent of Irish emigrants were female, while among Italians the percentage was a low twenty-one. The changing economic structure in Ireland shattered the entire community, women as well as men. Not only were jobs such as weavers, cabinetmakers, and shoemakers being eliminated by the indus-

The Cashman family sailed from Queenstown (Cobh), County Cork, at the height of the potato famine. (Harper's Monthly, Sept. 1884)

trial revolution; women, whose position has been compared to that of slaves, were servants and laborers of the lowest sort. In other words, in a collapsing society, practically everyone became a victim, women as well as men. Many of the female emigrants were young and single, determined that with any luck, a foreign country might mean more freedom, possibly a job, and maybe even a husband.[6]

Other factors aiding in the ability of the Irish people to escape such a debilitating situation were relatively cheap ship transportation costs, the lack of any immigration quota in such destinations as Canada and the United States, and the cooperation of American companies and the British government, both pleased to see more Irishmen leave Ireland; the Americans wanted the Irish labor, and the British government hoped partially to solve the "Irish problem" by having fewer Irishmen in Ireland.

Such circumstances meant that even poverty-stricken Irishmen could leave for the New World. For a few dollars, a ride in a cheap ship could be arranged for North America. This was one of the factors to the advantage of Irish women. The added distance from Scandinavia, Italy, or the German principalities meant longer journeys, and more cash.

The three Cashmans settled in Boston, which quickly became, along with New York City, the main focus of Irish settlement in the United States. The family probably arrived in the early 1850s, and they remained in Boston until 1865. Their trail is well established by 1864, when "Frances Cashman, widow," was living in a house at 328 Federal. The only known income for the family was from Nellie's job as an operator of a lift "in a public building." Her few comments about the work suggest that she liked the position. Such work brought her in contact with a wide cross-section of her fellow human beings, which would be an earmark of Nellie's environment for most of the next century.[7]

The next stop for the Cashmans was San Francisco, and motivations for the move are not known. In 1865 or 1866 the family left Boston by ship, traveled across the Isthmus of Panama by the newly-completed railroad, and completed the journey by ship to San Francisco.[8]

The attraction of California at this time was great for Irish-Americans. The Gold Rush was past, but much other history was still in the making. Over half of the Irish immigrants to the New World remained in the East, especially in New York, Massachusetts, and Pennsylvania. But California had a magnetism. In the Golden State, few traditions held sway. There was no rabid anti-Catholic movement, and the Irishmen, being English speakers, had a leg up on the other thousands of immigrants from Latin America, the Orient, and Europe.

By 1870, for example, San Francisco was the tenth largest city in the United States, and half of the population was foreign born. Within a few more years, one-third of the entire city was "Irish." This "Irishification" of San Francisco is not so well known as it should be. For example, New York City had an Irish mayor in 1880, and Boston elected one in 1884. Yet, Frank McCoppin, an Irishman all the way, had been elected mayor of San Francisco in 1867. California Senator John Conness, elected in 1863, had been born in Galway; Senator Eugene Casserly, elected in 1869, was from County Westmeath.[9]

These were far from isolated events. Saloon owner and politician Chris Buckley took his Irish machine to power in San Francisco in the 1870s and held sway into the next century.[10] By the late 1860s, when the three Cashman women settled in San Francisco, the city was becoming a paradise for Irishmen. Catholic churches and schools mushroomed, and city political and merchant offices were adorned with the nameplates and business signs of O'Malley, Sweeney, O'Meara, Donahue, Sullivan, Phelan, Casserly, and Murphy. Irish Catholic newspapers and magazines helped give this Hibernian community a sense of pride and accomplishment.

Soon after the Cashmans arrived in San Francisco they met Thomas J. Cunningham, another young Irish immigrant. They were all boarding in the same house, at 336 Fifth Street.[11] Nellie would be intertwined with the Cunninghams for the next half century, so some background is necessary here.

Young Frances Cashman and Tom Cunningham fell in love and were married in 1870. Perhaps this was one of the dreams-come-true of the Cashman women. Leave the decrepit conditions of Ireland for a new land, get to a place with a future, find a husband of like background and bright hope. Within the coming decade the Cunninghams had four children, who would become intimately associated with their Aunt Nellie.

Tom Cunningham had been born in Ireland in 1847 and was naturalized in San Francisco in 1867, when he was listed as a bootmaker. This remained his occupation, and he gradually improved his status. By 1874 he was secretary of the United

Workingmen's Cooperative Boot & Shoe Co., and over the years he was variously listed as a president, crimper, bootmaker, and superintendent. This was a small boot and shoe business but it would become important to Nellie, as San Francisco's twenty-two boot-shoe factories were the leading producers of mining footwear in the West. Nellie became familiar with her brother-in-law's products and would feature them when she went into business later.[12]

Boots usually meant mining in this era in the West, and one can only speculate on the type of conversation in the Cunningham-Cashman household. There was always some mining activity in the West, and Nellie obviously became closely attuned to it, by conversation with merchants, and by reading the local and mining press. She would become a newspaper scavenger in the coming decades, an omnivorous reader and advertiser. She kept in touch by reading, and she let others know what she was doing by advertising.

Her first mining community venture was in Lincoln County, Nevada, at the new silver town of Pioche. This was in reality a district, and "Pioche" included the town of Pioche and the nearby mining, milling, and transportation centers of Panaca Flat and Bullionville. This mining camp in the southeastern part of the state was first worked in 1868 with a primitive smelter. The town, named for San Francisco financier and mine-backer F. L. A. Pioche, by 1870 was the most active mining center in the state. At its peak — in late 1872 — the population of the town and vicinity was about 6,000. Such instant mushrooming was common in precious metals boom towns in the West.

Nevada in this era was known as a major silver producer. The leading region was still Storey County, the location of the fabulous Comstock Lode at Virginia City. Pioche, in Lincoln County, was not a rival but by 1873 had substantial production. The following chart deals with tons of both gold and silver ore extracted, and of course most of it was silver ore:

COUNTY	TONS EXTRACTED
Storey	293,717
Lander	40,895
Lincoln	37,776
White Pine	18,415
Humboldt	6,233
Elko	5,073
Esmeralda	3,406
Nye	2,152[13]

During Nellie's career she was often said to have the magic touch, to get to a mining camp in the early stages, profit, be aware of a potential demise, then take the money and run. The Pioche scenario comes close to fitting into this category. In July of 1872, Nellie Cashman and her mother, the elderly Frances Cashman, took themselves to Pioche and opened a boarding house, which they bought from David Leahy. The boarding house was at Panaca Flat, a small milling camp about ten miles to the south and east of Pioche; her mother, Fanny, was listed as the manager.[14]

Pioche has been described by most writers in unflattering terms. It was violent, dirty, and in general an unattractive place. As a mining camp it attracted hordes of young, single men who worked in the mines, mills, and smelters. Its peak years were 1870-1873. The vicinity provided these miners with relaxation in seventy-two saloons and thirty-two brothels, and the whole community was described by historian James Hulse as "a slovenly place."[15]

Some measure of the filthy nature of these communities can be had by viewing the newspapers of the era, which continually urged cleaning campaigns. For example, the *Pioche Daily Record* of October 4, 1872, reported on:

> the unsightly condition of our streets, caused by throwing refuse of all descriptions until the walkway as you walk down Main demoralized by packing cases, old boots, rags, discarded clothing, old wrapping paper and litter of every description meets the eye on every side.

Pioche, in eastern Nevada, years after the boom of the early 1870s which attracted Nellie and her mother. (Natural History Museum, Los Angeles)

This town was so violent that comparisons with other Western towns are not easily made. Only Bodie, California, had the same type of crude, uncaring, random and planned violence. Saloon brawls, town fires, street duels, and downright ugly, frequent confrontations were commonplace. Within a three-year period, there were at least forty violent deaths, mostly by gunshot. Tombstone, Arizona, by contrast, was an oasis of peace and tranquility.[16]

A catalog of vicious crimes, including gun fights, saloon brawls, senseless stabbings, and frequent theft can easily be made by reading the daily press and following the court calendar for Lincoln County. On one day, for example, the newspaper reported "A Fatal Shooting Affray," another shooting where Miss

Kate St. Clair got hit in the leg by a pistol shot, a further story entitled "Shooting in the Streets," and finally an article pleading for the hiring of more police officers.[17] The *Record* of January 26, 1873, when describing a bold robbery, opened the story with "Crime is rampant in Pioche."

One of the main ingredients in the cauldron of violence was the Irishman, who would be found throughout the Western mining camps. In fact, in one bitter gunfight in Pioche, David Neagle shot Jim Levy in the jaw; Neagle's father was from Ireland, and Levy was born near Dublin. This was the "initiation" of both men, who went on to famous careers; Neagle became Police Chief of Tombstone and a Deputy U.S. Marshal, and Levy became one of the most feared and respected gunfighters in the West. Besides these men, the Pioche "shooter's gallery" contained a large number of participants named Murphy, O'Grady, Shannon, Sullivan, Casserly, and so forth.

There were so many Irishmen in Lincoln County, and violence was so rife, that in many instances only Irishmen were involved. In November of 1872 there was a stabbing in the Headquarters Saloon, next to Nellie's boarding house in Panaca Flat. James Woods didn't like the way Bill McCarty was playing faro, so he put a large knife into him. Woods was from Ireland, and McCarty was a "Boston Irish." In March of 1873 in Pioche's Mint Saloon, Tom O'Brien pulled his pistol and shot Barney Murray in the chest. O'Brien then shot it out with a few peace officers before calling it a day. Both of these miners were born in Ireland.[18]

Pioche also suffered from that other frontier destroyer, the town fire. The town was practically obliterated by a fire storm on September 15, 1871. The area was so isolated that the first shipment of supplies for the devastated community arrived thirty days after the fire. When the town was rebuilt, shortly before Nellie arrived, no street or community planning was done, and the streets zigzagged, and businesses, mills, and homes were intermingled without much thought to civic order.[19]

The Lincoln County towns of Pioche, Bullionville, and Panaca had grown quickly, but in odd ways. For example, Pioche had 4,000 inhabitants, and there were several thousand scattered in the other nearby communities. Yet, there was no true hotel, not even in Pioche, although several establishments had "hotel" as part of their name. Comforts were provided by lodging houses and restaurants, which charged between $1 and $2 for an evening; meals in restaurants were between fifty cents and $1. Workers and residents of Pioche and vicinity sought survival, not the higher lifestyles.[20]

There were cultural and social institutions in Pioche that tried to alleviate the harshness of the mining-milling environment. There were Episcopal and Roman Catholic churches, a public school with over 100 pupils, and a county hospital. There were also celebrations and parades, and of course St. Patrick's Day was a major even in Pioche. There was also a very active St. Vincent de Paul Society which sponsored dances and other affairs to aid the local hospital.[21]

The Cashman name does not at first appear linked with the various charitable and religious activities in greater Pioche in the beginning. It may be that Nellie and her mother, just getting their start in business, felt that they had to concentrate their efforts on the boarding house and restaurant. However, towards the end of her time in Pioche Nellie was very much a part of the Roman Catholic and Irish scene, which would characterize her world for the next fifty years. In early October of 1873 Father D. Monteverde coordinated a "Ladies Fair" to raise funds for church causes. The week-long event featured sales of clothing, refreshments, gasoline stoves, and "fancy goods." Nellie Cashman and a Miss Kehoe were to handle the fancy cake with rose decorations, but they added functions as the week progressed. When the final tally was made, Nellie's table had taken in $389 from "refreshments and cigars."[22]

Starting in autumn of 1872 there was an additional spurt of activity at Panaca Flat, as the mines and mills had increased their output. One visiting reporter referred to it as a "beehive," with hoisting works, mines, mills, boarding houses, and saloons doing much business.[23]

Such activity led to changes in the Cashman operation. Up to this time the business was known as the name Miner's Boarding House, and Fanny Cashmen [the spelling in the advertisements] was the only name listed. This was Nellie's aged mother, and apparently being over seventy years old was too much to handle; or, it may be that Nellie had given her mother's name prominence out of respect.

In any case, in April of 1873 the advertisements began to read, "Boarding House . . . Miss N. Cashman, proprietress." She announced major changes, as she could now accommodate more boarders, and her house had "an enviable reputation for No. 1 board." Her place was at the head of Main Street in Panaca Flat, adjacent to the ice cream parlor of John Connerton, who supposedly had the only ice cream machine in the West. In her ad, Nellie announced that she had expanded, and would provide good board at low rates.[24]

As is common in gold and silver camps, the good ore goes quickly, and rare is the camp that had as much as a good ten years. The Pioche rise and fall can be seen in this production chart:

1870	$1,629,339
1871	3,592,151
1872	5,462,572
1873	3,709,864
1874	1,594,673
1875	1,010,133
1876	821,561[25]

Obviously, things changed after 1872, and as the months progressed, fewer miners were needed, not so many mill hands were kept on, the smelters could do with fewer workers, and so

BOARDING HOUSE.

IMMEDIATELY ABOVE CONNERTON'S
ICE HOUSE, PANACA FLAT.

MISS N. CASHMAN,

Proprietress.

THE PROPRIETRESS OF THIS HOUSE having recently added to and materially improved the building for the accommodation of Boarders, respectfully returns thanks for past liberal patronage, and solicits an increase of Boarders, as she is prepared to accommodate all who desire

Good Board at Low Rates.

The Table will be supplied with the best to be had in the Market.

apl-1m N. CASHMAN.

Miner's Boarding House

Panaca Flat,

HAVING FITTED IN GOOD STYLE THIS old established boarding house, the undersigned would be pleased to meet her friends.

☞ BOARD BY THE DAY OR WEEK. ☜

jy21-tf FANNY CASHMEN.

When Nellie and her mother Fanny first went to Pioche in 1872, Fanny was listed as being in charge. When improvements were made in 1873, Nellie was the manager. (Pioche Daily Record) forth. The wave effect was immediate: as fewer whiskeys were sold, not so many faro tables operated, and the gaudy don't-worry-about-tomorrow attitude gave way to reality. For those perceptive few, it was time either to leave or at least to look elsewhere for possibilities.

There were other signs of trouble in Pioche. By September of 1873 one mining reporter described the town as "quite dull at present." Much of the dullness was caused by bitter wrangling over ownership of titles and claims to mining property, and as a result some of the richer locations were idle. There was always the danger that men would be attracted to other mining camps, and this indeed did happen in Pioche. The silver mines in the Panamint Mountains of California began to get a big press notice, and many Pioche people started in that direction.[26] Because of these problems, unemployment was high, and few supplies, drinks, or meals were being purchased by men who could no longer afford them.[27]

Near the end of 1873, Nellie and her mother decided the time was at hand to do business elsewhere, or at least to get out of Pioche while one still could. They stopped running advertisements for their boarding house in early August, as they had decided that the flush times of Lincoln County were past. On October 17, they sold the Cashman Boarding House to Jacobs & Sultan, a firm that stayed active in the Pioche area for a few more years.[28]

The week after Nellie and her mother pulled up stakes in Pioche the local newspaper had several articles hinting of further troubles ahead. The *Record* admitted that "times dull — for Pioche." Yet, it urged folks to stay; after all, things may be duller elsewhere. Individuals and merchants, though, were selling out, and wagon-loads of furniture could be had cheaply. The local Justice Court was selling all sorts of goods to pay bills of insolvent debtors.[29]

Whatever the Cashman mother-daughter team earned during this brief foray in Pioche is not known. Because of Nellie's later reputation as a sound businesswoman, and as one who operated clean, decent places, she may have gotten her start earning this reputation in Pioche. She may have earned a good living while there, maybe even saved some money.

One thing is obvious. This was Nellie's first experience in a mining camp. She would live and work in at least twenty other mining camps and towns before she died, so most things about the Pioche experience must have been positive. After all, there were, even in that era, ways to earn a living other than selling bed and meals next to saloons, whorehouses, and filthy streets. The excitement of Pioche, the potential of making much money quickly, became a part of Nellie's personality, one that she would carry through four countries over the next fifty years.

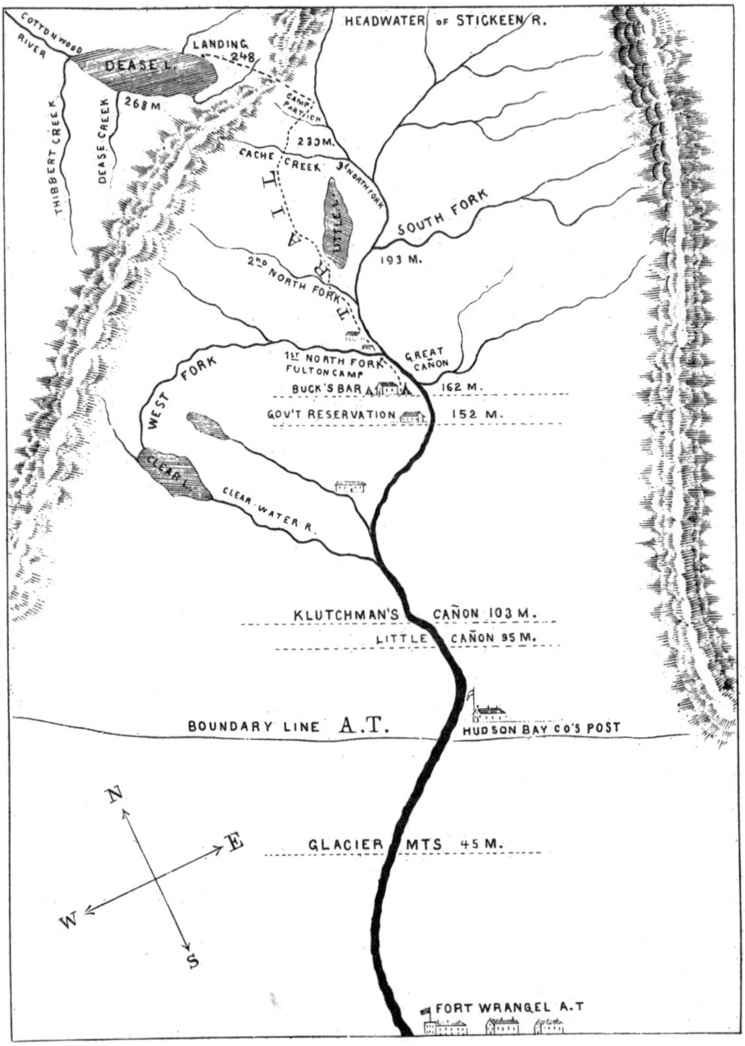

Map of Stickeen River and Cassair mines.
(Mining & Scientific Press, April 18, 1874)

II

CANADIAN CHALLENGE

THE Cassiar District is in the extreme northwest corner of British Columbia, a few miles from Yukon Territory, and the city of Juneau is directly to the West. The Alaskan towns of Sitka and Wrangell are also to the west and south of the Cassiar, which create the correct impression that this is indeed northern country.

Such geography would become associated and constantly repeated during Nellie Cashman's career. When she was in the British Columbia mining district she spent considerable time provisioning and traveling through Alaska. Among Alaskan female pioneers, Nellie is in the front rank. Decades before there was a Klondike Gold Rush, before the Golden Sands of Nome were discovered, Nellie Cashman had been trekking the wilds of the northland.[1]

British Columbia's first brush with gold came in the Fraser River excitement of 1858. This location is not far from modern Vancouver, which did not exist at the time. The small outpost of Ft. Victoria, on Vancouver Island, was the staging area for the miners and provisioners for the Fraser River mines. Thousands of men were lured to the Fraser River diggings from the California Mother Lode counties, but the rich material and the excitement lasted only a few years.

The land which would attract Nellie Cashman, the Cassiar

Mountain Range, was drained by the Stikine [Stikeen] River, a long, narrow body of rapidly rushing water that entered the ocean slightly north of Wrangell, Alaska. From the mouth of the Stikine up to the Cassiar Mountains is a very rough, mountainous terrain of several hundred miles. The Stikine River is more than 700 miles north of Victoria, truly in northern country.

The man who opened up this country was a talented, aggressive French Canadian named Alexander "Buck" Choquette. As a young man he participated in the California Gold Rush, went north for the Fraser River strike, and eventually settled at Ft. Victoria. His interest was in the northern country, where few had traveled.[2]

Choquette went north with a few partners in June of 1861, became friendly with the Stikine Indians (he would eventually marry one), and traded and explored up the Stikine. The *Victoria Daily British Colonist* kept its readers informed of Choquette's movements in late 1861, when it was feared he had been lost or killed by the Indians. The issue of January 10, 1862, announced that Choquette had survived, had found gold, and returned with $40 in dust a few days earlier. Some boat travel was possible at least 120 miles up the river, and Choquette worked with a knife and crude rocker in the sands and gravel.

The news caused a stir in Victoria, and a minor rush took place, and over the next four or five years, approximately 300 men prospected and mined up the Stikine River. One of the leading centers was at Buck's Bar, about thirty miles up the Stikine, which was later the Hudson's Bay Company trading post, where Choquette served as agent for a few years. This first "rush" into the area had really tapered off by the late 1860s.[3]

The major Cassiar discovery came a few years later, by another French Canadian, who had much different geography and a completely different agenda. Henri Thibert was born in 1843 near Montreal, headed for Minnesota and the Red River country in the late 1860s, and in 1869 set off on a cross-country exploration trip to Fort Edmonton, Great Slave Lake, and the Athabaska country. He linked up with another traveler named McCullough and they headed southwest, crossing into British Columbia near the modern town of Cassiar.[4]

They hunted and prospected for weeks, and in September of 1872, at the west side of Dease Lake, their fortunes changed. While McCullough was gathering white fish, Thibert explored a creek to the west that fed into Dease Lake. He found considerable gold on what would become one of the richest sites in the district, Thibert Creek. By the spring of 1873 United States Customs officials at Wrangell heard of the gold strike, and the news spread rapidly throughout the Pacific Coast ports.[5]

How, when, and where Nellie Cashman learned of the gold in the Cassiar District is not known. It may be possible that when she returned to San Francisco she read news reports of the placer fields in British Columbia. San Francisco had three vigorous daily newspapers, and was also the location of the West's leading professional journal, the *Mining & Scientific Press*. She was an avid newspaper and magazine reader and would certainly have come across many stories about the new gold workings.

However, there is more evidence that even in the distant deserts of eastern Nevada, Nellie learned much about what was going on in British Columbia. For example, in March of 1873 in Pioche there were at least two "Stickeen River" meetings, where the miners discussed the newest information from British Columbia, and there were also efforts in the planning stages to plan expeditions. There were subsequent meetings, and Nellie may have attended, or discussed the meetings with her many mining customers. Later, in the Cassiar District, she was frequently referred to as "Pioche Nellie," and quite a few men from Pioche did eventually go to the Cassiar. Their inspiration, as well as that of Nellie, may have come from attending these Stikine River meetings. For example, the *Pioche Daily Record* of January 14, 1874, in an article entitled "Going — Gone," reported that three parties had just left, one for San Juan, Colorado, another for French Guiana, and the third for the Stikine River.[6]

The details vary, but it is well established that it was in San

Francisco that she made up her mind to go to British Columbia. Some accounts state that she was joined by 200 Nevada miners who wanted to go north, an obvious exaggeration. One of her interviews with Fred Lockley seems to be a balanced account:

> In 1874 I went to San Francisco where I joined a party of six adventurous spirits who were outfitting to go to the mines. Some of them wanted to go to South Africa, some wanted to go to the newly discovered mines in British Columbia. We tossed up a coin, heads for South Africa, tails for British Columbia. It fell tails up so we went North.[7]

We must remember that Nellie didn't just happen to be in San Francisco. She and her mother had left Pioche, Nevada, to return home, where her sister Fanny and family lived. It was also in 1874 that Nellie went to Edouart & Cobb photographic studio on Kearny Street, where she had the only known photo of her youth taken; she was thirty at the time.[8]

Nellie left San Francisco in the spring of 1874, sailing for Victoria. As it would turn out, 1874 was to be the height of the Cassiar Rush, and Nellie was to participate in its rewards. She and the few men in her traveling party most likely had loaded up on mining equipment and clothing in San Francisco. Nellie's brother-in-law, Cunningham, specialized in miners' boots. Victoria, though, was where they completed their provisioning, such as the purchase of foods and blankets. The party then took a steamer up the inside passage to Wrangell, Alaska, then to the Stikine River, where by small boat they traveled up to Telegraph Creek, arriving in early summer.[9]

In this era, the word Stikine rather than Cassiar was most frequently used to describe the new mining district. The San Francisco *Mining & Scientific Press* ran frequent articles on the

Never a large city, Wrangell, Alaska, was on the Inside Passage and was a major military and supply point for both southeastern Alaska and northwestern British Columbia. (Natural History Museum, Los Angeles) ▶

rush, giving all sorts of advice on travel conditions, use of sleds and toboggans, lack of paths and roadways in the area, and so forth. The issue of March 28, 1874, by an unidentified writer, is a good summary of the type of information being published:

> I left San Francisco on the 5th inst. and had a very quick trip to this place, being a little over three days — distance 800 miles; price of passage, cabin, $30; and steerage, $15. There is only one boat running from 'Frisco to Victoria, but there are three boats running from here to Fort Wrangell at the mouth of Stickeen river, which is 800 miles from Victoria; — fare $30, cabin; and $20, steerage. From Fort Wrangell up the Stickeen river to Buck's Bar, it is called 160 miles; from Buck's Bar across the country to the mines it is 80, and some say 100 miles. There are no provisions nor supplies at the mines yet, nor will there be till after the river opens . . . . As near as I can find out, there are 500 or 600 men already in there, and on the way. The only supplies they can get are what they take with them on handsleds.

Therefore, incoming parties had to be well-provisioned, and they faced a mountainous country, whose only avenues were a partially navigable river and practically unusable trails. This was the situation when Nellie and her party arrived at the mouth of the Stikine River, within weeks of when the above account was written.[10] Another consideration was that two jurisdictions were involved in getting to the gold fields. Ft. Wrangell was in Alaska, that is, in United States territory. The mouth of the Stickine River was also in Alaska. Yet, fifteen miles above the mouth one entered British Columbia. So, this gold rush in Canadian land, primarily by people from the United States, meant that the stampeders provisioned in Canada (Victoria), went north and debarked in the United States (Ft. Wrangell), then went up the Stikine River where they again entered Canadian territory.

By early summer Nellie had reached Dease Lake, a thin body of water about twenty-four miles long. At the north end of Dease Lake, to the west, was Thibert Creek, where Henri Thibert had first discovered gold. Nellie, and most of the prospectors and miners of 1874, were at mid-point on the lake, at the west side. Here, where Dease Creek emptied into Dease Lake, a small town emerged, named variously Laketon, Laketown, or Lake Town.[11]

For this mining summer, about 1,500 men were in the vicinity, with as many as 400 to 500 around Dease Creek and Dease Lake. Snow drifts and ice hung on until late spring, and gradually a tent city was built, with a few stores and saloons. Soon some log structures were erected, including Nellie's combination boarding house/saloon.[12]

The mining camp at Pioche had been one of silver ore, where huge quantities of rock were mined, crushed, then processed to reveal the silver. Pioche had been a camp of large operators and mill complexes, which left little room for individual initiative. There is no evidence that at Pioche Nellie was involved in mining in any way.

However, the Stikine-Cassiar diggings enabled her to develop a pattern that she would use for half a century. This was a gold placer area, which referred to gold nuggets and dust which had been removed from rock matrix by erosion and weathering over the millions of years and transported by water to this site. Placer or alluvial mining in its initial stages was ideal for the individual, and a prospector or miner with the most elementary tools and equipment could work an area for profit. Only when gold is found in quartz and other ores, where large machinery is needed, is the individual at a disadvantage.

Nellie opened a boarding house and earned her income from bed, board, and booze. On the other hand, she also learned about certain promising claims and purchased them when the time was right. Furthermore, she got involved in "grubstaking." Having confidence in certain men, or in their description of their properties, she provided them with money and supplies. If they hit, she also profited. In one of her interviews she stated clearly, "I alternately mined and kept a boarding-house for the miners."[13]

The summer of 1874, then, was spent by Nellie in a furious pace to acquire and work claims, maintain a decent boarding house which brought in additional money, and purchase new supplies for those just up the hard trail, or those who were "packing it in" and heading back to civilization. "Miss Nellie Cashman" according to one journalist, "was one of the few white women who reached Cassiar last year, where she opened a boarding house on Dease Creek and realized a comfortable "pile."[14]

Nellie left the diggings in October and headed for Fort Wrangell. There she boarded the "S. S. Californian" and headed south, getting as far as Nanaimo, on Vancouver Island, where a severe onset of fog stopped the vessel. However, excitement and the pride in announcing their great gold victories couldn't hold back the party. The *Daily British Colonist* of November 3, 1874, announced:

> Miss Nellie Cashman, and a few others, arrived in canoes from Nanaimo, yesterday, having reached that port on Saturday afternoon last, in the S. S. Californian from Ft. Wrangel . . . . $300,000 in gold dust from Cassiar.

The $300,000, of course, referred to all the gold on the steamer, but Nellie at least had a small portion of that impressive amount.

Nellie by this time was well enough known to get her name mentioned occasionally in the press. She was unusual in that she had been in two mining camps, where usually the only women were prostitutes or hangers-on. Yet, she was an attractive, intelligent young lady, in business, without a hint of scandal about her. What happened next, though, would propel her into the limelight in a fashion that would be repeated many times in the years to come.

Within weeks after arriving in Victoria in early November of 1874, Nellie heard of a potential disaster in the Cassiar diggings. The onslaught of a severe winter had cut incoming supply trains, and stopped hundreds of miners from leaving the area. They were trapped, without provisions, and scurvy was said to be rampant.

Nellie at once hired six men to help her, and she bought provisions, especially lime juice and other foods thought to alleviate scurvy. They took a steamer back to Fort Wrangell, arriving in mid or late December, and headed up the Stikine trails. They used snow shoes, and because the snow was so soft and deep they could not use dogs. Each of the seven in the party pulled a sled. The challenge of more than 100 miles in rugged terrain in such weather is difficult to appreciate. On some days they made as little as five miles, for their burden consisted of 1,500 pounds of supplies. Nellie used two blankets to keep warm at night. At one point, while she was sleeping at night, an avalanche buried Nellie, and when her mates came looking for her she was just digging out.

The party spent most of January and February on the trail. Sometime in mid-February, a lonely mail expressman from Dease Lake heading towards Fort Wrangell, reported the following:

> On the day following, he met the famous Nellie Cashman and five men. The woman was on snowshoes and was as

> jolly as a sandbuoy. At the Boundary Post, she lost the trail, and was 28 hours exposed to the pitiless pelting of a storm, without shelter or blankets.[15]

They made the trip to Dease Lake, arriving in late February or early March, and Nellie was so preoccupied with nursing back the men to better health that she postponed mining, and could not open her hotel for the season on time. This rescue trip by Nellie became part of Northern folklore, and in the later camps at the Klondike and Nome, mention of Nellie and Cassiar brought a knowing nod, referring to the unselfish expedition into a certain death to help her fellow miners.[16]

Because it has become folklore, a dozen or so authors have written about the trip, with nary a sentence based on what really happened. Some details of the dangers and exertions came from later interviews with Nellie. However, the event did indeed occur. One of the standard contemporary notices of her famous trip appeared in the *Daily British Colonist* of February 5, 1875, and contains the type of prose, the kind of assessment that would become part of Nellie's reputation:

> Her extraordinary freak of attempting to reach the diggings in midwinter and in the face of dangers and obstacles which appalled even the stout-hearted Fannin and thrice drove him back to Wrangel for shelter is attributed by her friends to insanity. So impressed with this idea was the Commander at Fort Wrangel that he sent out a guard of soldiers to bring her back. The guard found her encamped on the ice of Stickeen cooking her evening meal by the heat of a wood fire and humming a lively air. So happy, contented and comfortable did she appear that the "boys in blue" sat down and took tea at her invitation, and returned with her.

Nellie spent much of 1875 and 1876 at the Dease Creek camp, mining, running her boarding house, and doing rather well. Nellie also displayed other characteristics for which she would become well known in the coming decades. On one trip from

Cassiar to Victoria she sent her aged mother in San Francisco $500 in gold. To Nellie, relatives, friends, and strangers could all line up for physical and mental comfort. In the summer of 1875 she made her way through many of the outlying camps in the Dease Lake area, gathering contributions. While in Victoria she had become familiar with the Sisters of St. Ann and learned that the new hospital of St. Joseph's was under construction. Through her prodding, the Cassiar miners donated $543 to the cause.[17]

The rich placer areas were quickly worked out on the Cassiar, and although the mining population in the vicinity stayed at over 1,000 from 1875 through 1879, gold production fell noticeably:

1874	$1,000,000
1875	830,000
1876	556,474
1877	499,830[18]

This was not the type of activity for Nellie, as decline in production meant a smaller number of boarders, fewer meals, poorer claims to work, and much less chance to make money by grubstaking a prospector. In May of 1876 Nellie wound up her business at Dease Creek and in a more pleasant, almost casual way, made her way down the Stikine river and trails to Fort Wrangell. There she boarded the Hudson's Bay Company's "S.S. Otter" and headed for Victoria, arriving there on June 15.[19]

By mid-summer of 1876, while back with her mother and married sister in San Francisco, Nellie could add up her profits from the gold fields and look back on two widely divergent mining camp experiences. She had spent time in the desert silver mining camp at Pioche, Nevada, and in the placer gold diggings of northern British Columbia. She had moved in communities of masculine values and population and emerged with a reputation that would see her through countless mining camps in three countries: she was an attractive young lady who was aggressive, caring, charitable, and unafraid of challenges.

III

ARIZONA CALLS

ARIZONA did call, but Nellie has not enlightened posterity about when she got the call, nor why. After her Cassiar venture, she spent approximately a year in San Francisco, most likely studying prospects for finding a lively Western camp. She determined that Tucson, Arizona Territory, would be her base. At that time, Tucson and mining had nothing in common, so the reader may be justified in wondering about the attraction of the relatively small settlement.

Nellie was basically a "boomer," a "stampeder," which meant in the West of the late nineteenth century one who was attracted to and was early on the scene in new communities. In most cases these new places were mining and milling camps. However, railroading was another great creator of boom towns, and places such as Denver, Cheyenne, and Laramie owe their origin and rapid growth to railroad events. Some settlements, like Fairbanks, were both mining and railroad centers. Thousands of Westerners were part of this "boom" mentality, which is why Nellie over the years would meet so many acquaintances in her travels: the Earps, Dave Neagle, Jim Levy, John Clum, Arizona Charley, and many others.

The railroad situation in the American Southwest at that time consisted of a broad plan of the Southern Pacific Railway to head east from California and link up at Deming, New Mexico,

Tucson in the early 1880s; this was Nellie's first place of business in the Southwest. (Harper's Monthly, March, 1883)

with the westward moving Atchison, Topeka, and Santa Fe Railroad. By 1877 the Southern Pacific had opened traffic between Los Angles and Yuma, Arizona Territory, and was laying track towards Tucson. Events went on schedule, and the ceremonial spike was driven on the opening day of service to Tucson on March 17, 1880.[1] The *Daily Arizona Star* of May 1, 1879, had stated that the railroad would "connect Tucson with the outside world by a band of iron." Because of such developments, "wealth and prosperity were inevitable."[2]

Apparently Nellie was convinced that Tucson, formerly a sleepy Spanish and Mexican pueblo, would evolve into a major transportation center. She headed for Tucson, via Los Angeles. The "City of Angels" made an impression on her. "There was nothing, absolutely nothing but dogs in sight. It was the first time I knew that dogs had souls."[3] Nellie traveled by the new Southern Pacific line into Yuma, then took the stage to the coming metropolis of Tucson, where on October 10, 1878, she was listed in the "recent arrival" column of the *Arizona Weekly Star*.[4]

Nellie had decided that what Tucson needed was a restaurant, so she made lease arrangements, began tidying up a building on the south side of the Church Plaza, and in June of 1879 opened the Delmonico Restaurant. The advertisement promised "The Best Meals in the City," as Miss Cashman "has had long experience in the business and will personally superintend the cooking and dining apartments."[5]

The advertisement was arranged by Nellie visiting the office of this new newspaper, which was owned by John Clum. He was much taken with her, as "she was the first of her sex to embark solo in a business enterprise." Clum also stated years later that "her frank manner, her self-reliant spirit, and her emphatic and fascinating Celtic brogue impressed me very much, and indicated that she was a woman of strong character and marked individuality."[6]

The Tucson of this era "improves on acquaintance," according to a traveling correspondent of the *Chicago Tribune*, who visited the pueblo in mid-1879. He used such words as "dilapidated and unprepossessing," but believed that a transition was in process. The houses were one-storey adobe style, weather-worn, with unpainted doors. The Catholic church was "rough-looking externally," and the streets were without sidewalks, not yet lighted, but liberally supplied "with Mexican dogs and curs of high and low degree."

But, evidence of busy trade was everywhere, with the regional mining districts as well as with the Mexican state of Sonora. Several large hotels, saloons, and restaurants were getting busier, and that sign of frontier progress — gambling — was in evidence.[7] Institutional changes were also in evidence. The local Presbyteri-

DELMONICO

Restaurant.

Miss Nellie Cashman

Has just opened a New Restaurant on the South Side of Church Plaza,

TUCSON, ARIZONA.

The Best Meals in the City

Can be obtained here. The tables are supplied with every thing the market affords.

Mrs. Cashman has had long experience in the business and will personally superintend the cooking and dining apartments.

PATRONAGE SOLICITED.

Come and satisfy yourself on the truth of this announcement.

South side of Church Plaza.

(Daily Arizona Citizen, August 14, 1879)

ans had organized in 1876, the Methodists in 1879, and the Episcopalians in 1880. The Roman Catholics, of course, had been in this part of the Southwest since the 1700s. An ice plant was opened in 1879, and St. Joseph's Hospital was built and opened in 1880.[8] In other words, when Nellie began her restaurant business in Tucson in mid-1879, the town was shaking off its sleep and beginning to take advantage of all sorts of mercantile possibilities, due largely to the railroad developments and increasing mining activity, in both Arizona Territory and Sonora.

Ed Schieffelin, a true desert rat, caused Nellie Cashman and around ten thousand other people to shift attention to southeastern Arizona. Schieffelin had wandered several Western states and territories looking for mineral wealth. In the summer of 1877 he picked up an interesting piece of silver "float" and filed his claim in Tucson, the seat of Pima County, on September 1, 1877. He named it "Tombstone," referring to what a few U.S. soldiers had told him was all he would find in the vicinity.[9]

What became "Tombstone" consisted of zero population in the summer of 1877. By 1880-81, there were approximately 5,000 people living in the new community, with a few thousand scattered in surrounding mine and mill areas, such as Contention, Charleston, and Millville. This was one of the more startling, eruptive examples of community building on the mining frontier.[10]

Silver does not exist free in nature, as it quickly combines with other elements. As a result, it is very difficult to identify, and once found, presents many milling and smelting problems. Almost all ores of silver were found at Tombstone: argentite, a silver sulphide; cerargyrite, or horn silver; cerussite, a silver-lead combination; and galena, proustite, and bromyrite. Chemicals, crushers, refiners of many types, and smelters are needed in order to arrive at "bullion," bars or cakes of raw silver that are then ready for jewelry, silverware, industrial use, or any of a multiple of uses for this precious metal.[11]

From the beginning, therefore, at Tombstone the large operators were in control, as heavy financing from San Francisco, Philadelphia, New York and other cities was needed to purchase the expensive equipment and hire engineers, foremen, and craftsmen.

While rapid deployment of mills and smelters was taking place, the establishment of what would become Tombstone evolved. A townsite association was formed on March 5, 1879, and a site near the location of Toughnut Mine was selected, but only after controversy. The first house was erected in June of 1879, the first hotel at the same time, and in December the village of Tombstone held its first election, electing William Harwood as its first mayor. A stage line operated by J. B. Kinnear was in

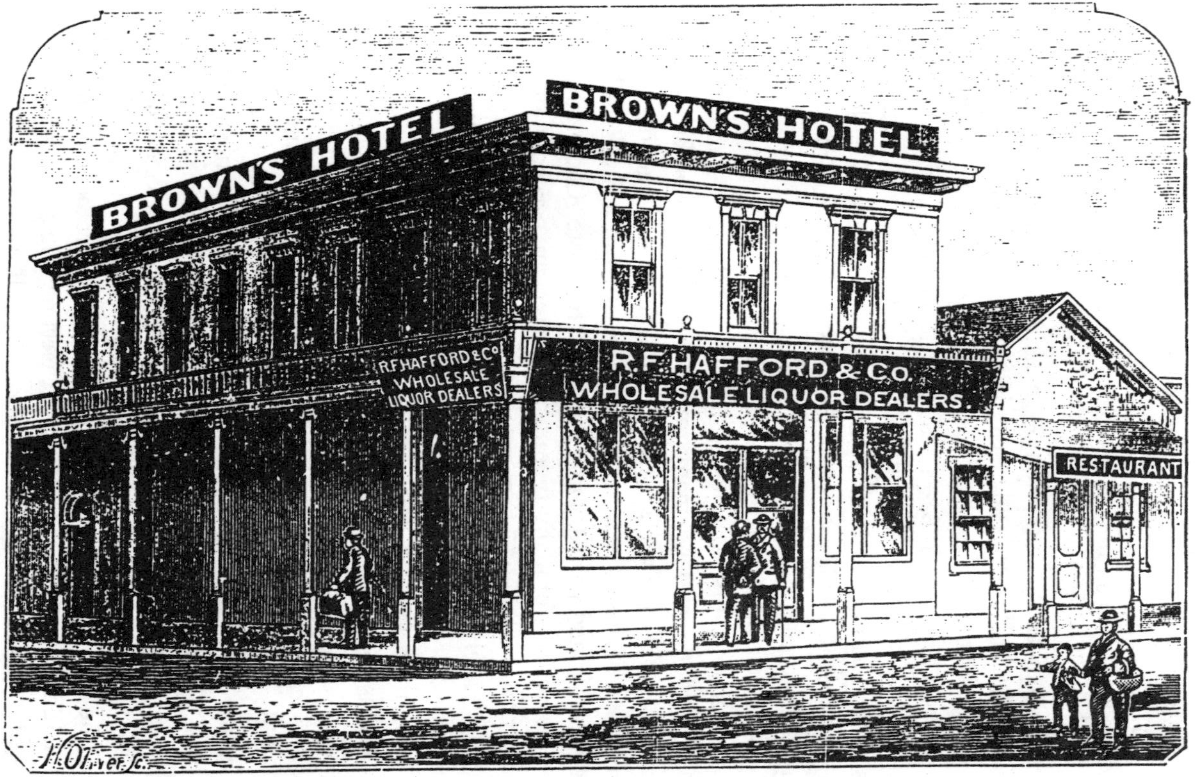

Tombstone, nonexistent in 1877, was by 1880 a vigorous city of several thousand inhabitants. (Arizona Quarterly Illustrated, October 1880)

operation, and the *Tombstone Nugget* issued its first number in October of 1879.[12]

All of these events transpired in less than two years, an indication of the enthusiasm and optimism many people felt for this new camp, which quickly became an incorporated city by early 1881. One of its many drawbacks was lack of transportation. The railroad by 1880 reached Tucson, but Tombstone was more than 75 miles to the southeast. Although the railroad tracks continued to be laid eastward, Tombstone in its glory years, 1879-85, was without rail service.

One of the oddities of Western frontier life was the rapidity in which "town histories" were published. The Tucson *Daily Arizona Citizen* of February 17, 1880, brought its readers up to date on the rise of the new mineral haven to the southeast. Hotels and restaurants were plentiful, there were many good restaurants, "saloons are numerous," and there were seven physicians and ten lawyers busy in town. A school was in session, and "grocery and dry goods stores are well supplied with heavy stocks." There was no church yet, "nor any prospects of one." The author was pleased with the residents, most of them "old miners who have roamed over the whole Pacific Coast for the last thirty years in search of the precious metals."

The emergence of Tombstone from zero population to boomtown enthusiasm was read about and heard by certain types throughout the West. U.S. Deputy Marshal Virgil Earp in Prescott, the Arizona territorial capital, put out the call for the Earp clan. Wyatt and James from Dodge City, Morgan from the copper mining town of Butte, Montana, and Warren from the family home in Temescal, California, joined Virgil in arriving in Tombstone in the early throes of the boom. Gunfighter-miner-lawman Dave Neagle heard about the town while in San Francisco, recuperating from a mining accident; he immediately left for Tombstone. John Clum, who had just established the *Citizen* in Tucson believed that the new town of Tombstone needed another newspaper, and he went there and established the appropriately named *Epitaph* on May 1, 1880.

In a large mining-milling community much of the real profit went to the operators of hotels, saloons and gambling halls, and to the merchants. Nellie Cashman was as aware as other stampeders of the beckoning call of Tombstone. It was true that Tucson was increasing in importance, but why not go the center, the cause of that new importance? The throbbing pulse of the mineral activity in the Southwest was in Tombstone, so Nellie had little choice in the matter.

Nellie decided to keep her Delmonico Restaurant in Tucson operating temporarily, but she headed for Tombstone and quickly arranged to enter business. The *Tombstone Nugget* of April 1, 1880, announced her plans, with strong complimentary language:

> Miss Nellie Cashman, of the famous Delmonico restaurant, in Tucson, has opened a gent's furnishings goods store in Tombstone, on Allen street, adjoining Ward's market. She will keep a large supply of furnishing goods, both for the ladies and gentlemen, including boots and shoes, and as Nellie was never outdone in any business she has undertaken, her success in our midst is double sure. Call at her place of business.

One could logically ask why a restaurant operator would move to a new community and open a gent's furnishing goods store. A mining-milling economy meant men, and far fewer women and children. Furthermore, Nellie knew about boots and shoes, and she had a prime source in San Francisco: her brother-in-law Tom Cunningham, of the United Workingmen's Shoe Company. An article in the *Citizen* of May 12, 1880, also explained that she had a business partner, Miss Kate O'Hara.

These rapid events contain hints of Nellie's career in the coming decades. She had very few preconceptions about what it took to be successful. She was confident enough that she could swing from operating a boarding house to running a restaurant, to selling boots and shoes, or to getting down on her hands and knees and panning for gold. Also, her taking up with Kate O'Hara showed another of her methods of operating. Sometimes, for ready cash or for help in running an establishment, she linked up with other men and women. She did what she needed to do in different circumstances. Also, she had no intention of maintaining a shrinking business, or staying in a dying community.

In a business directory published in the new community, it was stated that Nellie's store was on "Allen Street, above Fifth," and that she had opened for business on April 15. In advertisements for the subsequent few weeks, Nellie pushed the phrase "THE NEVADA BOOT & SHOE STORE." She was using every angle she could. The emphasis on boots was the result of the contact with her brother-in-law, and the "Nevada" was leaning on some of her prior mining camp experience, when she had spent two years in Pioche, Nevada.[13]

The swiftness with which Nellie worked, the rapidity of business decisions is evident from her moves in June of 1880. Nellie and Kate O'Hara parted, but on friendly terms. Kate returned to San Francisco, and will be heard from shortly. Nellie then linked up with Jennie Swift, also from Tucson, and opened the Tombstone Cash Store, at her same location on Allen Street. Groceries, provisions of all kinds, and fresh fruits were promised daily from Los Angeles suppliers. The store was owned by "Cashman & Swift, Proprietors." Not only were advertisements placed in both Tombstone newspapers, but the Tucson *Arizona Daily Star* mentioned the new venture, and "we wish these ladies unlimited success in their new home and business place, as they really deserve Fortune's smiles."[14] The store shelves must have been crowded, because a news item from the same period mentioned that Nellie had just arrived from a trip to Tucson, back at her Nevada boot and shoe store.[15]

Tombstone was a wide-open town, but not in the sense portrayed by traditional movies or modern television. This was a bustling place, where miners, merchants, mine owners, entertainers, saloon-keepers, and various professionals and visitors thrived in

THE TOMBSTONE

Cash Store,

ALLEN STREET, NEAR FIFTH.

CASHMAN & SWIFT, PROPRIETORS.

Groceries & Provisions,

OF ALL KINDS.

Fresh Fruits Received Daily

FROM LOS ANGELES, CAL.

Call and See our Stock.

When first arriving in Tombstone, Nellie opened a boot and shoe store, and shortly later added her "cash store." (Arizona Quarterly Illustrated, July 1880; Tombstone Nugget, June 24, 1880)

the midst of what was nearly pure desert. Stores, hotels, schools, physician's, and lawyer's offices provided the residents with a fare and services as good as could be obtained in any American city. Tombstone may have been a bit more expensive [most things were imported from Los Angeles and San Francisco], but a steady flow of cash in this boom town made good living possible.

On the other hand, Tombstone was not a wide-open town in the sense that it was a center of frontier violence. In its short, turbulent heyday, two major factions did emerge: the mining community, most merchants, most Northeners, and the *Epitaph* on the one side, and the "cowboy" faction that lived in the outlying districts, most Southerners and Texans, and the *Nugget* on

THE NEVADA

BOOT AND SHOE STORE

AND GENERAL

FURNISHING GOODS EMPORIUM

Where every article for Gents' Ladies' and Children in the above Line, can always be fouud.

—FINE F LAVORED--

CIGARS & TOBACCO

GENTS' HATS AND DUSTERS,

ALLEN STREET,

BET. FIFTH & SIXTH, *TOMBSTONE*

MISS NELLIE CASHMAN.

the other. This rivalry and bitterness did lead to the famous Tombstone shootout of October 26, 1881, when the Earps obliterated the Clanton cowboy faction. But this was an aberration in a usually law-abiding community.

To understand this point clearly, one has only to consider the Earps in Tombstone. Virgil, Wyatt, Morgan, Warren, and James lived there during most of 1880, 1881, and part of 1882. All, except James, held various law enforcement positions, with Virgil being the most important, as Deputy U.S. Marshal and Chief of Police. Yet, during all this time, none of these Earps fired a shot in Tombstone except on the day of the shootout. The only exception to this was once when a wild, drunken young Warren Earp fired his pistol into the air and was fined. Historians who are familiar with violent communities in the West do not rank Tombstone high on the list.

Obviously, in a community with two factions, you had a choice of siding with one, or trying to work with both. Nellie's commitments, her cause, was with the mining community, not with the cowboy faction. However, she and several other merchants used obvious good sense and advertised in both newspapers, even though her great admirer, John Clum, was the mayor and publisher of the *Epitaph*.

Though in the middle of this desert in the American Southwest, we must refer again to Ireland. A new county had been created in 1881 following the Tombstone discovery; Cochise County, with Tombstone as its seat. A subsequent census determined that about 9,600 people lived in the county, and of these 5,300 lived in Tombstone.[16] The following partial list indicates the national origin of the Tombstone residents:

Americans	2880
Irish	559
Mexicans	423
Germany	300
England	279
Canada	254
China	245

The remainder consisted of small numbers of Norwegians, Russians, Italians, French, Turks, Austrians, and so forth.

Notice that the Irish contingent was highest of the foreign-born. This, of course, was a bit misleading as hundreds of the "Americans" were also of Irish origin, men such as Cochise County Sheriff Johnny Behan.

The *Epitaph* announced these figures on July 8, 1882, and concluded with this statement:

> The Irish are the least clannish of the foreigners, and mix indiscriminately with other peoples, while the English almost invariably live together.

Think of what this meant for Nellie and her various businesses. Because of her experience and common sense, she would

Tombstone's mayor, John Clum, championed Nellie's causes there and would later meet her in the Klondike. (Arizona Quarterly Illustrated, January 1881)

have done well as a merchant. Yet, she was in a new community where the largest majority were Irish-born, who got along very well with others, and who were involved in mining, a field where Nellie had already established a reputation in Nevada and British Columbia. Truly Nellie Cashman was in an ideal setting.

More personnel reshuffling followed almost as soon as Nellie went into the grocery business. Kate O'Hara had previously bowed out and left for San Francisco. On July 15, 1880, the *Nugget* announced that Nellie had purchased the interest of her partner Jennie Swift, in both the boot and shoe business and the grocery establishment. The newspaper praised Nellie's "sterling business qualities," and suggested that by doing business strictly for cash she could sell at lower prices than some of her rivals.

In the following months there were frequent notices and advertisements in the Tombstone newspapers to inform the readers of what was new with Nellie's various enterprises. She was "receiving daily fresh fruits and vegetables," could always be called on for "groceries and provisions of all kinds," received frequent shipments of Nevada boots and shoes, and in late October was said to feature "fresh roll butter" as well as rosy-cheeked peaches.[17]

The Tombstone merchants were able to provide the finest foods, alcohol, and clothing to their customers in spite of having only stage connections to the world. When Nellie was in Tucson, most of her restaurant provisioning had been from Los Angeles dealers. For her Tombstone boot and shoe operation and for grocery business she dealt with wholesalers from San Francisco, Tucson, and Los Angeles.

In San Francisco, Nellie used her former partner Kate O'Hara as a buyer for several years. Records show the following as regular business contacts for Nellie: G. W. Gibson of Los Angeles, groceries; Woodhead & Gay, Los Angeles, fruits and seeds; E. Germain, Los Angeles, wholesale produce; I. Hanley, San Francisco, teas, coffee, spices, yeast powders; Michelssen Brown & Co., San Francisco, pork packers; A. Vignolo, Los Angeles, groceries.[18]

The business whirlwind continued for Nellie, though, as she tried to find her niche in Tombstone. Around October 20, 1880, Nellie opened the Arcade Restaurant & Chop House, which adjoined her Cash Store. The Arcade had been operated by Joseph Pascholy, with whom Nellie struck up a lasting personal and business relationship. The *Nugget* ran several notices about her plans, and on October 31 the first advertisement on the Arcade was published:

> Again at my old occupation. Having had years of experience in the restaurant business I know whereof I speak, that the Arcade will serve Better Meals than any House in Town.

She definitely kept up her boot and shoe business, as advertisements continued to run for months. And, she probably also retained her grocery, fruits, and vegetable Cash Store, because the products would also have been used in the restaurant.[19]

A minor change occurred around this time in her overall arrangements. The Arcade building, which had been owned by Thomas Devere, was bought out by Joseph Pascholy on November 16. The purchase included Nellie's lease.[20] Pascholy, who earlier couldn't make a go of the Arcade, now decided to buy out the entire building. Pascholy was a cattle man with a ranch in the Dragoon Mountains, and he and Nellie may have discussed this new arrangement. Nellie Cashman and Joseph Pascholy would be in business together several times in the next few years.

Although Nellie had a reputation as a good manager and was known for efficiency and cleanliness in her places, she was far from a drudge. The theatre and schools did not seem to attract her attention, but she became instrumental in pushing two needed institutions in Tombstone, a hospital and a church. This was a heavily male-dominated community, and Nellie believed the miners needed places to repair their bodies and their souls.

She would be associated with Catholic church plans in most

communities where she lived. John Clum's *Epitaph* of September 25, 1880, carried this notice:

> Nellie Cashman, the irrepressible, started out yesterday to raise funds for the building of a Catholic church. We don't know what success attended her first effort but will bet that there is a Catholic church in Tombstone before many days if Nellie has to build it herself.

What Clum didn't know was that the plans were already being converted into action. Nellie even arranged for some of the lumber to be hauled in from the Huachucha Mountains for the church and rectory, which were erected in late 1880-early 1881, and she may also have provided the lumber for the parish hall.[21]

The following announcement appeared in the *Nugget* of October 7 and is a good summary of what was done and who participated:

> I take this method of extending thanks to the citizens of Tombstone for their liberality in subscribing to the fund for the building of a Catholic church in this city. Also, to many kind friends in Charleston and Contention City, who responded so generously to this most worthy cause. I can state that $700 has been raised, and all the bills for lumber and other building material being in, we know very nearly the exact cause of the construction of the church, say $1,000. It is proposed to give an entertainment soon, for the purpose of increasing our present fund, and as a liberal patronage is always extended in a good cause in this community we are not doubtful of the result. Again I extend my most heartfelt thanks to the public generally for their liberal support.
>
> NELLIE CASHMAN

Nellie, obviously, had even corralled mill workers from nearby Contention and Charleston for donations. This was in keeping with her wide-ranging geography. In the mining district of Cassiar, she had dunned miners to contribute to the new hospital in Victoria, 800 miles away.

The role of Nellie Cashman in the planning and funding for the Catholic Church and for the new county hospital has been mentioned in many sources, but in a fragmentary way, which suggests bursts of energy on her part. In reality, she was the lynch-pin of both movements. She was part of the planning-funding drives from the beginning, and she was on committees for both the church and the hospital to see that cash was raised and that other commitments were kept.

The *Epitaph* of September 26, 1880, printed what the reality was when an article headed "Catholic Church and Hospital" was opened with the sentence, "Miss Nellie Cashman informs us that she and her able adjutant, Mrs. Twomey, have collected in cash about $300 more in subscriptions toward the building of a Catholic Church in Tombstone." Nellie's places of business were used for both meetings and for the depositing of contributions for church and hospital.

Her work with the church was better known than her hospital assistance, but both took much of her energy and time. On the hospital group she was the treasurer, was on the committee for bylaws and constitutions, and she and two other members

Proposals for Building a Church.

Sealed proposals will be received at the store of Miss Nellie Cashman, on Allen street, up to 7 o'clock p. m., Monday, October 4th, to build a Catholic Curch on the lot northwest corner Safford and Sixth streets. The building will be 24x40 feet and 14 feet high. The ground needs grading about 10 to 14 inches. The plan and specifications can be seen at the office of J. Reilly. Proposals should be endorsed, "Proposals for building Church," and addressed, "Trustees Catholic Association." The Association will provide all the material. By order of the Trustees.

JAMES REILLY,
Chairman pro tem.

were in charge of selecting a site for the hospital. One of her associates on the church and hospital committees was Justice James Reilly, one of the most intelligent, cantankerous, well-known men-about-town. Nellie was known to express her opinions clearly and frequently, so meetings with her and Justice Reilly in the same room must have been exciting.

The Catholic church plan was effective, money was obtained, and a building erected. The Sacred Heart Church of Tombstone held its first services on November 28, 1880, with Rev. Father Antonio on hand. The first baptism was held on February 18, 1881.

The hospital plan for Tombstone did not meet with such success. Although by late 1880 there were at least seven physicians practicing in Tombstone, they did so out of their homes or offices. A meeting was held at Leo Goldschmidt's store on October 18, where a Miners' Hospital Association was reorganized. Officers and trustees included such people as W. J. Hunsaker, an attorney, J. A. Kelly, owner of Kelly's Wine House, and John L. Fonck, a partner in Goldschmidt's furniture store. Nellie Cashman was the only woman to receive recognition, and she was appointed to the key position of treasurer. A county hospital began to function early in 1881, and on a few occasions Nellie would be involved in finding religious personnel to serve as nurses on the staff. She was in frequent contact with the Sisters of St. Joseph in Tucson and San Diego, and they would provide much of the nursing expertise for Cochise County.[22]

Events related to Irish pride, and Ireland's needs, continually got Nellie's attention. When Schieffelin Hall was inaugurated on March 17, 1881, the first event was celebration of St. Patrick's Day, with an Irish League Ball. On June 17, 1881, the local branch of the Redpath League held a Grand Ball, as a fund raising effort for the Irish Land League. There were four different committees for the events, with a roster of names like Neagle, Behan, Kelly, Costello, Fitzpatrick, Rafferty, and Murphy. The main committee, the Committee of Arrangements, included Miss Nellie Cashman.[23]

In this spring Nellie may have dropped her Nevada Boot & Shoe business, as advertisements no longer appeared. She continued running the Arcade Restaurant, and it was described as being in the Vizina Block with meals at all hours, regular dinner from four to seven p.m.[24] She closed the doors at eight p.m. The street map of Tombstone indicates that Nellie's Arcade was in the heart of the bustling community. She was in the same block as the Safford & Hudson's Bank, and right down the street was Milt Joyce's Oriental Saloon, where Wyatt Earp owned part of the action and dealt faro on occasion.

That Nellie had action in mind, not merely talk or plans, can be seen in the frequent lists of "freight arrivals" published in the Tombstone newespapers. Benson, Arizona, was the point where the trains from California and Tucson deposited their freight, which then had to be transported to Tombstone by wagon. The *Epitaph* of March 11, 1881, listed more than thirty merchants receiving packages of goods, such as Spangerberg [gunsmith], one package; J. A. Kelly [saloon], twelve packages; Blinn [lumber dealer], 1 car lumber, and so forth. On this list, "Miss N. Cashman" was listed as receiving forty-one packages for her various Tombstone enterprises.

In late spring of 1881, then, in this lively community Nellie was running the Arcade Restaurant and was in the thick of such local activity as helping the Irish Land League and gathering funds for the Catholic church. Satisfied? Apparently not. On June 8 the following notice appeared in the *Nugget*:

> Having leased the Arcade Restaurant to Mr. I. A. Korka, I hereby announce that I have retired from business. It is important that all outstanding accounts be settled immediately. Thanking my old friends for past favors, I hope they will bestow their patronage upon my successor.
>
> Tombstone, June 1, 1881 NELLIE CASHMAN[25]

Nellie was not pessimistic about Tombstone's future. Rather, she was already looking elsewhere. About forty miles to the south in Cochise County, practically on the Mexican border, was the young copper mining town of Bisbee. Discovery and first workings of this magnificent copper deposit occurred in 1880. This would become one of the world's leading copper centers but was yet in its formative stages.

The Arizona newspapers for a few months whipped up considerable enthusiasm for the Bisbee copper district, and Nellie, an avid reader, apparently fell under the spell. For example, on March 15, 1881, in a detailed article, the Tombstone *Epitaph* had as an opening line, "Bisbee is now enjoying a genuine boom." Words like "success," "boom," "untold riches in ore," and similar mining camp hype were music to any stampeder's ears.

Nellie, of course, liked to be in a mining camp in its youthful stages so had made queries and a visit and was convinced she could do well there. She leased the Bisbee Hotel [sometimes called the Bisbee House] and planned to open there sometime in July.[26]

On June 22, three weeks after Nellie had disposed of her Arcade Restaurant, Tombstone was engulfed in flames. The fire started next door in the Oriental Saloon, where several men with lighted cigars were measuring the contents of a liquor barrel. An explosion occurred, and one half of the business section of Tombstone was soon engulfed in flames. The Vizina Block was severely damaged, and Nellie's former Arcade Restaurant was reduced to ashes, as were forty or more other buildings in Tombstone.[27]

Nellie spent the few weeks following the fire traveling between Tombstone, Tucson, and Bisbee, trying to get all things in order for her new Bisbee operation. In early July, at her own expense, she went to Tucson and returned to Tombstone with three Sisters of Mercy, who were to take charge of the new temporary city and county hospital.[28]

By mid-July Nellie was in Bisbee, operating the Bisbee Hotel. John Clum's *Epitaph* of July 8, 1881, gave her a good sendoff, stated that she could keep a hotel better than the majority of men, and regretted losing "one of its best citizens, and Bisbee gains an ornament to its active business interests."

The promise for Bisbee was the future, not the present. Tombstone, in spite of its devastating fire, quickly grew from the ashes, and throbbed and thrived at once. Bisbee, even farther removed from transportation centers, was on nobody's highway. Furthermore, the requirement for the huge machinery needed for copper milling and processing meant that several years would pass before that copper camp was in full swing. Nellie's timing had been faulty, and the customers didn't exist who should have been knocking on her hotel and restaurant door. Within a few months of opening, Nellie closed and returned to Tombstone. And not quietly. The owner of the Bisbee House, Mary Christy, brought action against Nellie for unpaid rent in the Cochise County justice court.[29]

During her time in Arizona Territory Nellie had been supporting her aged mother in San Francisco, sending periodic checks for $50. In 1881 an event happened which would alter Nellie's life. Her brother-in-law in San Francisco, Tom Cunningham, died on February 20, age thirty-nine. He left behind Nellie's sister Fanny and five young children. For Nellie the obvious followed, and by late summer Fanny and her five children arrived in Tombstone, the beginning of "Aunt Nellie's" lifelong commitment to the young Cunningham family.[30]

Nellie's sister Fanny was not a freeloader. In September the sisters arranged for a lease on the Vickers building on the north side of Fremont Street, between Fourth and Fifth. They had some repairs and carpenter work done, the building was painted, and new beds and other furniture were purchased. The Delmonico Lodging House was opened for business in early October of 1881, "and will be kept cleanly, in the best of style." The proprietor was Mrs. T. J. Cunningham.[31]

Although the Cashman sisters were working like dynamos to get the Delmonico open, this was just one of Nellie's around-the-clock activities. At the same time Nellie was getting her own

NELLIE CASHMAN
Russ House Sunday Bill Of Fare
October 30, 1881

SOUPS

Chicken | Consomme

FISH

Brook Trout | Fois a l 'huile

BOILED

Lamb, Caper Sauce
Beef al'Espanol
Corned Beef

ENTREES

Breast of Lamb, breaded a la Mayonnaise, Croquettes of Rice, Kirshwasser Sauce, Chicken Fricasse a la Creme, Salim of Chicken Giblets, Calf Head in Tortue.

ROASTS

Prime Beef, Ribs of Beef, Leg of Mutton Stuffed Lamb, Dressed Veal, Pork With Appleasuce, Chicken

VEGETABLES

Sugar Peas, Tomatos, Corn, Turnips, Mashed Potatoes

PASTRY

Assorted Pies and Jelly Roll

PUDDING

New York Plum, Heart Sauce, Lemon

DESSERT

Grapes and Walnuts

SALADS

Lobster, Tomatoes, beets and Horseradish

RELISHES

Assorted

(Archives, Sisters of St. Ann, Victoria)

business in order. She joined with Joseph Pascholy, who had earlier owned and leased the Arcade Restaurant to Nellie. The Russ House, a combination hotel and restaurant at the corner of Fifth and Tough Nut, was up for sale and Pascholy took it over. The *Epitaph* of September 30 announced that the Russ House would open that evening "in a first-class style," and breakfast the next day would be its first meal. After sizeable investments of time and money, the Russ House had its first large meal on October 2, 1881, with "four hundred persons partaking of the good things set before them."[32]

The Russ House had been built in December of 1880 but had never lived up to its promise. Nellie and partner Pascholy intended to provide every reasonable convenience to their guests. Nellie's menus became well-known, and were frequently printed in the two Tombstone newspapers. In addition to large, clean bedrooms, attention was given to other comforts: W. A. Caddy was put in charge of the bar. For the next six years, off and on, Nellie would be associated with the Russ House, sometimes in partnership with Pascholy. His job was mostly as financier, and the meat for the table was from his ranch in the Dragoon Mountains.[33]

Opening of the Russ House and the Delmonico Lodging House did not mean that Nellie was too busy for her causes. Near the end of October 1881 there was a flurry of moves when the local branch of the Irish National Land League pressed its Arizona fund drive. Nellie was the treasurer, and forwarded to the national office $200 that she had gathered. Tombstone did its share, but Tucson, in spite of having about 1,000 Irishmen, did very little. The *Epitaph* of October 29 asked if all Irish patriotism had been "burned out" of the Tucson residents. Probably not. What Tucson was missing was Nellie Cashman. Her vigorous anti-English and pro-Irish prodding was present in Tombstone, not Tucson, hence, the difference in "patriotism."[34]

By the end of November Nellie was engaged in more charitable work, heading up a raffle for "an elegant bedspread." The piece was created by the Sisters of St. Joseph of Tucson, and the proceeds from the raffle would be to supply "their personal wants."[35] Nellie had been close to this religious group while in Tucson, and would continue the relationship in the coming years.

Bedspreads and charities were not the only order of business in Tombstone at this time. On October 26 was the culmination of the deadly townsman-cowboy feud, when the Earp faction forced the hand of the Clanton and McLaury families in the famous shootout on Fremont Street. Three of the cowboys were killed, and although Virgil Earp was Chief of Police and believed he was merely enforcing the law, he was removed from office. Nellie Cashman may have been out of town on this occasion.

Nellie had other problems. The Russ House was doing very well, as was the Delmonico Lodging House. Unfortunately, her sister was not. Advertisements in both Tombstone newspapers in December listed Nellie Cashman's one-half interest in the Russ House for sale:

> It is to be sold on account of a sister's sickness, which necessitates my entire attention elsewhere. The house is doing a splendid business. Apply to MISS NELLIE CASHMAN, at the Russ House.[36]

Things went from dismal to almost disastrous, as the *Nugget* on December 31 carried an advertisement for the sale of the Delmonico Lodging House, "owing to the sickness of the proprietress, Mrs. T. J. Cunningham, sister of Miss Nellie Cashman."

Nellie believed she had no choice. With a seriously ill sister, whose five children were all under the age of ten, Nellie could operate neither her own place of business, nor that of her sister. By January both businesses were in other hands, and Nellie could only hope that her sister would recover.

Fanny Cunningham gradually improved, and Nellie was able to resume some of her favorite activities. For example, in March she was busy with another Irish National Land League charity drive and ball, and was coordinating affairs with the Sisters of St. Joseph of Tucson.[37]

One of the more reliable chroniclers of that era was John Pleasant Gray, and he mentioned that Nellie's various lodging houses in Tombstone often acted as a "free hospital," and that Nellie had more than influence with the miners, she often controlled their vote. Nellie also knew the where and who of charitable giving. According to Gray, Nellie frequently solicited for church and hospital funds in the red light district, where she was welcomed and where the contributions were important.[38]

When Fanny recovered from her illness, the sisters were ready for full-time business again, as both had drive, and certainly a need. Nellie's mother in San Francisco depended on her, and Fanny had five young children to clothe, feed, and educate. Nellie looked around Tombstone, found a building that seemed to her to be in the mainstream, and made arrangements:

> A new hotel is about to be opened in this city, which will be known as the American. Miss Nellie Cashman and Mrs. Cunningham are the proprietresses of this new house, and will conduct the same upon a plan highly satisfactory to the public. There will be attached a large and new dining hall, where a first class table will be spread, at which place elegant meals can be had at all hours. The American house is on Fremont street, between Fourth and Fifth.[39]

The sisters pushed work on the place and the first meals were served on the afternoon of April 25, 1882, and the *Epitaph* of that date described the "elegant and elaborate style," maintaining that customers would find there "as palatable meals as can be had in the city."

Off to such a good start with their new restaurant, Nellie and Fanny had one good month of business before the nemesis of the frontier struck again, a devastating fire on May 25 that gutted Tombstone's business district.

The fire started in a water closet at the rear of the Tivoli Saloon on Allen Street, caught in the framework of the building which was dry and "burned with the rapidity of cotton." Dozens of business houses were utterly demolished; for example the publishing offices of the *Nugget* on Fourth Street were described as "swept from the face of the earth."

Some merchants, such as Mrs. Fly at her boarding house, were able to fend off the flames, or to save valuable furniture or other material. The American Hotel, too, on the south side of Fremont, was in the path of fire, but more than luck saved the day:

> The American Hotel, kept by Miss Cashman and Mrs. Cunningham was in imminent danger. But the plucky ladies stationed several of their friends with buckets and kept the building thoroughly saturated with water, thereby preventing the flames from communicating.

The sisters did not come out of this disaster unscathed. Their entire wardrobe was lost, Nellie had a severe burn on her arm, and Fanny's hair was singed. Insurance covered most of their losses.

Nellie's close friend Milt Joyce, who owned the Oriental Saloon, had been hard hit during the great fire of 1881, but on this occasion his building suffered only minor damage. But Milt was businessman par-excellence, as "all the liquors he had on hand he distributed to the gallant defenders of the city," not making any enemies in the process.[40]

The American Hotel went back in business after a few weeks, and Tombstone gradually recovered from its second flaming holocaust. Yet, Tombstone the vital, growing mining and milling community was already in its mature years but was not aware of it. This period of 1882-83 was its peak, with a population of between 5,000 and 6,000, and more than a dozen major mines and mills operating in its immediate vicinity. No further major bodies of silver ore were located, and some of the working mines were finding that the depth of the ore bodies was not so extensive as they had hoped.

There were other mining and milling problems that were emerging, such as water in the mines, and difficulty in crushing and processing certain of the silver ores. Although the "slide" was not immediate, 1882 for Tombstone meant a leveling off, then the next few years would see a gradual diminishing of activity.

Nellie was not immune to these doings, and for the remainder of 1882 and 1883 she bided her time, keeping on the lookout for new adventures while at the same time earning a living with the no longer exciting American Hotel in a Tombstone recovering from the fire.

The silver mining town of Kingston, New Mexico, was not exciting enough for Nellie, and she stayed less than a year. (McKenna, Black Range Tales)

IV

WEAK LEGENDS, GREAT ADVENTURES

TOWARDS the end of 1882 and the beginning of 1883 Nellie Cashman and her sister Fanny Cunningham were pretty much out of business. As usual, Nellie had saved more than enough for a rainy day, and she was called upon during this period to bear the brunt of all the finances for the Cashman-Cunningham household. Nellie and all the Cunninghams were now living in a small adobe cottage behind the Russ House.

Fanny had been neither a malingerer, nor a thorn in Nellie's side. Fanny was a dedicated mother who worked hard, when she could. Unfortunately, she couldn't work often. Fanny was wracked by tuberculosis, and she was frequently in bed. By 1883 Fanny Cunningham was out of the picture due to a permanent state of sickness.

By this time Nellie had closely cultivated relationships with the Roman Catholic priests and sisters in Tombstone, Tucson, and San Diego. In the coming decade Nellie would farm out the Cunningham children to the close religious supervision of these people she knew so well. Mike Cunningham spent several years at the Catholic school in Tucson, and later attended St. Michael's College in Santa Fe and St. Vincent's College in Los Angeles. Several of his younger sisters boarded with the Sisters of the Home of Peace in San Diego.

Fanny, unable to care for herself, was sent to the Sisters of the Home of Peace sometime early in 1883. She may have died there, or may have returned to live with her mother shortly before her death. In any case she died on July 3, 1884, and was buried in Calvary Cemetery, San Francisco. Which means that from 1883 onward, Nellie Cashman was the sole support, the determiner of education for her five nieces and nephews. In the coming fifteen years Nellie would live in more than a dozen mining camps in the United States and Mexico, but she was in constant communication with these children, visited them when she could, and received regular reports from the people in whom she had absolute confidence, the priests and sisters of the schools, boarding houses, and convents.[1]

Spring of 1883 brought to Western North America news of another gold discovery, this one in Mexico, half way down the peninsula of Baja California. The supposedly rich placers were found between Calmalli and Mission Santa Gertrudis, a location just north of the current boundary between the states of Baja California Sur and Baja California Norte. This is about 450 miles south of the border at San Diego. However, there was no road to the region, and to this day there is no north-south railroad in the Peninsula.[2]

Santa Gertrudis was founded by the Jesuits in 1752, and the Dominicans erected a small stone church, completed in 1796. The vicinity was known to contain gold placers, and was even worked periodically. A mission report of 1863 referred to the ground as "auriferous," and suggested that important gold, copper, and iron could be worked in the vicinity.[3] Yet, nothing was done; no substantial discoveries were made until later.

Who discovered the gold placers and when is not clear, but newspapers in Guaymas, Sonora, and San Diego, California began carrying news of the riches in mid-April of 1883. By mid-May the news had become a rage, and extravagant accounts abounded.[4]

The placenames Mulege as well as Guaymas figure prominently in this gold rush, yet neither location was close to the diggings. Guaymas was a port in Sonora, across the Gulf of California from Baja California, but became the prime location for gathering of men and supplies for the gold fields. Mulege was a small fishing village and port on the Baja California coast, but was almost 200 miles south of the gold fields. Yet, it was the only location resembling a settlement where vessels from Guaymas could touch. From Mulege, some parties went north overland, while others went north by smaller vessel and debarked at Trinidad Bay, not too far from the diggings.[5]

There was another, dismal possibility for getting to the gold fields, but that was forced by necessity on miners coming from San Diego or Los Angeles. They traveled by steamer to the port of San Quentin, on the Pacific Ocean side of the Peninsula. Then the parties had to travel overland across the Peninsula, and down to the placers, over 150 miles of rugged, desert terrain. One newspaper account suggested that the placers were "within easy reach from San Diego via San Quentin," a masterpiece of optimism.[6]

After a few weeks of "rush," authorities got a better handle on the situation and records of individuals and ships began to be kept and were published. Between May 18 and May 23, for example, 78 passengers left Guaymas headed for the gold fields, identified as follows: Mexicans, 21; Americans, 51; Greeks, 3; English, 2; Irish, 1. Records were not so easily kept for the few overland parties from Los Angeles or San Diego. The 51 "Americans" listed above were almost all from Tombstone and Tucson.[7]

One of the earliest detailed accounts of the placers concerned the travels of Jacob Schreiber, who returned to Tucson around May 21st and was interviewed by the *Citizen*. Scheiber was in Mulege when the discovery became known, as several Indians and Mexicans came to town with coarse gold. Mulege became "depopulated," as the residents flocked to the gold fields. Schreiber joined a Mulege party and went to the area, where they found many Mexicans getting nuggets from the gravel. Water was scarce and had to be packed in from twelve miles away. "The excitement is drawing in people from all parts of California, and large parties are en route to the mines."[8]

Such reports received even greater credibility when Manuel Riveroll arrived in Mulege on May 17 with forty ounces of gold from the placers and praised the future of the camp. Riveroll was a prominent merchant who was known as a reliable conveyor of information, which only encouraged the speculators, prospectors, and miners.[9]

Obviously, with talk of these gold fields on the lips of every miner in Arizona, Nellie Cashman was not the last to hear of the news. The next month in her life can be fairly well documented, but if so one is justified in wondering at the contradictory prose generated by more than a dozen writers who have romanticized her time in Baja California. The truth is much more fascinating and reveals Nellie's character and reputation.

The legends, summarized briefly, suggest that Nellie and a band of miners grabbed a stagecoach in Tombstone, went down towards the gold fields, marched inland, and almost died of thirst. Nellie went on ahead to the Mission, got water, received blessings by the padre, saw rich gold placers, but decided not to mention them to her dying partners. She returned to them with the needed water, saved their lives, and they retreated to Arizona.[10]

Well, a stagecoach was taken, but only to the railroad station at Contention, west of Tombstone. In 1882 the railroad between Nogales to Guaymas was completed, the first United States-owned railroad in Mexico; earlier the portion of the railroad from Tucson to Nogales had been inaugurated. The entire line was part of the Southern Pacific Railroad system, although the name differed in Mexican territory. The railroad went from Nogales on to Magdalena, then to the capital city of Hermosillo, and finally to the port of Guaymas on the Gulf of California. In fact, Mayor Charles Strauss of Tucson and a party of twenty-two "notables" took a train ride to the "Sister Republic of Sonora" to Hermosillo, had a meal with Governor Luis Torres, then after a further three-hour train ride, arrived in Guaymas. This excur-

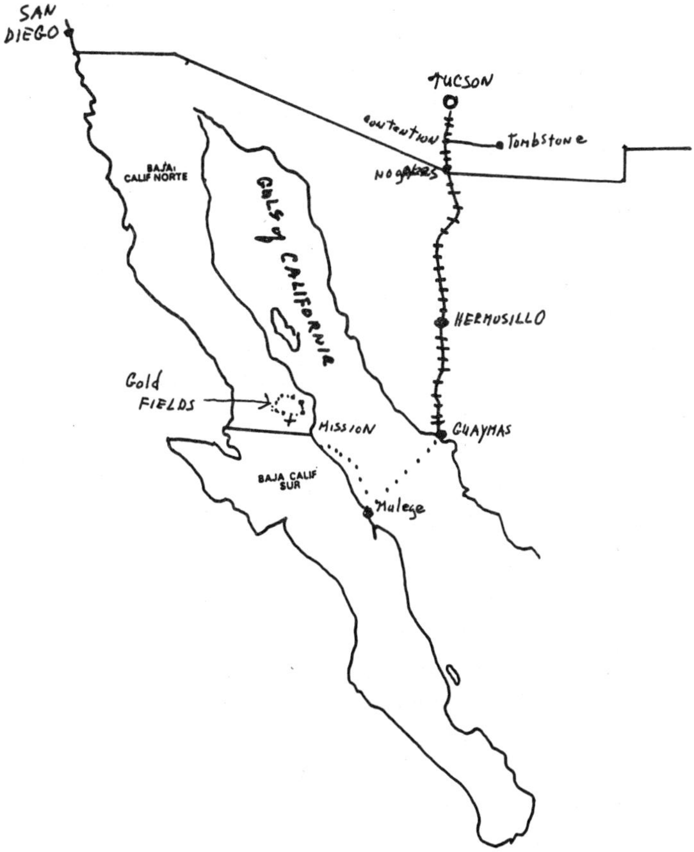

This misguided gold rush to Baja California led to near-tragedy for Nellie's party, but she enhanced her reputation for leadership in a man's world.

sion was completed on May 15, when the gold news was yet of a minor character.[11] The point is that to get from Tombstone or Tucson to the port of Guaymas was by then a relatively easy operation, to be carried off in less than a day.

The excitement in Tombstone matched that anywhere, and soon some of the leading citizens were planning to invade the Baja California placer fields. Nellie Cashman was in charge of the party, and whether or not it was her idea, or she was elected, is unknown. Nellie had been in mining camps in Nevada, British Columbia, and Arizona, and although her Nevada and Arizona stays had been in areas of silver ore, her years in the Cassiar diggings of British Columbia had been in working gold placers, and that was what they were seeking in Baja California.

Twenty-one stalwarts from Tombstone enlisted in Nellie's party, and although the names of all are not known, two of them were among Tombstone's leaders. Milt Joyce, owner of the Oriental Saloon, was also a Commissioner of Cochise County. Mark Smith, local attorney and well-known personality, who would later become a U. S. Senator, also enlisted under Nellie's charge. In some accounts the group was referred to as the "Joyce-Cashman Party," but probably because Joyce was so well known. Or, Joyce may have helped bankroll the expedition.

The group obtained provisions, and the twenty-one hopefuls set out on May 20 from Tombstone, and after a brief coach trip west to the rail station at Contention boarded the train; on May 21 they arrived in the port of Guaymas. Nellie had her typical working outfit on, which was black trousers, blue shirt, and high-top boots. She arranged the twenty-one men in a military-like organization, and each man was assigned a particular task, such as carrying the water, being responsible for maps and papers, making certain proper tools were obtained, and so forth. In the words of one journalist, "she assumed personal charge of her own party."[12]

At Guaymas, the Cashman party purchased some burros, and then sailed on May 23, most likely on the schooner "Gisna" and headed for Mulege. Ho! To the gold fields![13]

Unhappily for the party, the good times were behind them. The trip across the Gulf took almost four days to Mulege; then almost two days were consumed in traveling in a smaller ship one hundred miles up the coast of Baja California to Trinidad Bay. From here the party was faced with an overland hike of about 100 miles to the placer fields, over rough trails, lacking in water, and usually without a hint of vegetation.

Nellie decided that only part of the group should go on, test the gold fields, and then bring up the rest of the party. She and a group of five or six men, with burros, left Trinidad Bay and forged ahead. Very soon the intense desert heat and their rapidly disappearing water supply spelled potential disaster. They could not proceed. They stopped, boiled in the sun, and were about to become ex-miners.[14] The Arizona party were not tenderfoots, but the desert here was much harsher than any they had known. One report suggested that "lizzards do not exist there," and going ten miles was comparable to traveling fifty miles in Arizona. An Austrian and a Frenchman died of thirst a day or so after the Cashman party had set out.[15]

Their certain death was avoided by luck, but the details vary. One account has it that the rest of Nellie's party came up just in time to provide water for them. A different report had Mexicans returning from the gold fields, where they had purchased some gold from the Indians. On their return trip they came to the assistance of Nellie, "nearly dead from exhaustion on the first trip of sixteen miles." This fits in with the account from John Hart of Tucson, whose party was behind Nellie a few days but knew many of the details of their hardships. He maintained a diary of his trip and recorded:

> A Mexican, coming from the Mission of Santa Gertrudis, reported having met the rest of them lying famishing for water between the first water hole and Trinidad Bay. One of the Mexican storekeepers immediately sent them a pack mule with water.[16]

That was the end of the venture. However, it was more than desert hardship that turned the Cashman party back. They were saved, and if they then had heard glorious reports of huge nuggets, gold dust all over the region, they might have pushed onward. But by this time practically every report, every returning prospector and miner, had only negative words.

There had been gold there, but apparently not much. The general consensus was that Mexicans and Yaqui Indians had in the past year pretty much worked out the fields. There was some gold remaining in spots, but available water was between ten and twenty miles away, and the Indians, not being ignorant of needs, were charging fifteen cents a gallon for water. The John Hart party found water, but when they reached Mission Santa Gertrudis all they found was "a crowd of discouraged miners returning from the placer fields."[17]

Desert hardships coupled with such bad news about the placer fields was enough for the Cashman party, and they turned around and headed for Trinidad Bay. After a wait of a few days they obtained a vessel, and headed back to Guaymas.

By this time newspapers in San Diego, Tucson, Los Angeles, and San Francisco were running stories about "The Bubble Burst" and other warnings about the search for easy gold in Baja California. Apparently Nellie's party arrived back in Guaymas on June 9 or June 10. A news account from Guaymas on June 11 indicated that a party just arrived which had "almost succumbed to thirst" and found that the fields had already been worked out: "Miss Nellie Cashman, who had arrived at Guaymas on her return from the mines, confirms the above statement."[18]

The Cashman party immediately took the railroad back to Arizona, where at Contention on June 11 Nellie was interviewed by a correspondent "DIXMONT." She referred to the riches as a "delusion," and reported that she had obtained all that was possible in Baja California: cacti, rocks, and Spanish bayonet. Lack of water, high temperatures, and no gold, she pointed out, were reason enough for failure. In spite of that failure, Nellie impressed that reporter with her planning skills and conduct of the expedition. This was an example of "what an energetic and plucky woman can do."[19]

There were dozens of other small and large expeditions to the Baja California gold fields in that two month period, and few did very well there. However, it was the publicity of the Cashman party that spelled the end of the short-lived boom, that removed its luster. The newspapers of California, Arizona, and Sonora warned travelers in advance that not enough evidence was yet in to justify launching expeditions. The news in San Francisco, Tucson, and San Diego that even Nellie's well-organized group found only hard times there pretty much ended the 1883 Baja California gold rush.

In the midst of failure, Nellie Cashman received only praise. The fact that she organized an expedition, went around a thousand miles from Tombstone, and pushed into the desert wilderness in charge of a party of miners embellished her reputation. The San Diego *Union* of June 16, 1883, praised her efforts with this vigorous prose:

> In all the vicissitudes of life she has maintained the highest self respect, but is as ambitious in her notions as Joan of Arc. In Arizona she could raise a company at any time who would follow her to the death, either in search of gold or Apaches.

Faced with replenishing her purse after such a financial drain, Nellie went to work at once, and at the end of June took over the management of the Russ House in Tombstone.[20] However, the Baja California trip was more than a psychological blow to Nellie. The Los Angeles *Times* of July 18, 1883, reported that Nellie was in Tucson securing the services of Dr. J. C. Handy; her "health was seriously impaired by her expedition to the gold fields of Lower California."

Although the Tombstone area was a mining-milling region for large companies with complicated machinery and metal processing systems, that did not necessarily mean that individuals had no role. Nellie Cashman was not out in the hills panning silver [a most uneconomical effort for anyone]. However, she and other individuals did participate in prospecting, and more importantly, in the constant buying and selling of mining claims.

Sometimes a claim could be adjacent to a fairly rich operation, whose owners wished to expand. At other times, renewed examination of the claim would reveal ore missed on an earlier study. And, there were occasions when most companies refused to work certain ore because it was not rich enough, but other companies felt that a different process or new machinery could wrest dollars from the ore.

In her years in Tombstone Nellie bought and sold many claims and made some money in the process. Probably her earliest action was to locate a claim in the Chiricahua Mountains, which she named the Parnell Mine, in honor of the great Irish political hero Charles Stewart Parnell. In 1882 she obtained the Big Blue claim in the Huachucha Mountains, and in the following year bought the Last Chance in the Turquoise Mining District, the Littlefield and New Year in Cochise District, and part of the Big Comet near Tombstone.

She bought and sold more claims in 1884, received $700 for her part of the Comet claims from Malcolm McQuade. Nellie continued to buy and sell claims, including one named Nellie, near Tombstone, and other named Chasman by the Alpha Mine; this later was copyist's error, and the claim owner was recorded as "Nellie Chasman."

In many of these ventures, Nellie acted in partnership rather than alone, but when she did so her name was usually placed first on the legal documents. This did not indicate politeness on the part of her partners or the notary publics; in most cases the business arrangements occurred because of her initiative.

Often her partners were Irish, and this could be by personal preference, or because there was such a heavy Irish contingent in Cochise County. Some of her partners were James Hensley, Thomas T. Kavanaugh, Malcolm McQuade, and Dennis McCarty.

Several mining deals of 1883 indicate Nellie's confidence and obviously comfortable financial status. She bought the Last Chance and the New Year claims on April 16 for a total of $603, before she left on the Baja California venture, and on August 29 purchased the Big Comet claim for $700; this, of course, was after her return from Mexico. In other words, disaster that the Mexican trip was, it obviously did not wipe out Nellie's savings.[21]

For most of 1883 through 1886 Nellie earned a steady income from operating the Russ House, sometimes alone, sometimes in an arrangement with Joe Pascholy. Even though Tombstone had leveled off and was beginning to slide as a boom community, Nellie's reputation, and the good location of the Russ House, assured her of a decent living.

Two events of 1884 that are so closely associated with Nellie's "Angel" career have to be considered, even though her role in these affairs has never been established.

In early December of 1883 the "Bisbee Massacre" occurred. Five men entered the community on horseback and tried to rob the Castaneda & Goldwater store. The holdup was planned by John Heith, who had come to town weeks earlier. The robbery was ill-planned, and shooting erupted in many directions. Two Bisbee residents were killed outright, a man and pregnant woman were to die in hours, and another innocent bystander was severely wounded. The outlaws escaped, having received a few dollars for their efforts and the hatred of everyone in southern Arizona Territory.

Eventually the outlaws were captured in various parts of the Territory as well as in New Mexico; they were sent to Tombstone for safe-keeping and trial. John Heith, the ringleader, was tried separately, found guilty, but for only second-degree murder. He received a life term in the Territorial Prison at Yuma.

Such a sentence would not be tolerated by citizens of Bisbee or Tombstone, and on the morning of February 22, 1884, Heith was taken from the jail in Tombstone and hanged. Nothing was done to the conspirators. In fact, Dr. George Goodfellow came up with one of his classic statements, saying that Heith "died of Emphysema of the lungs which might have been caused by strangulation self-inflicted or otherwise."[22]

Such a fate was not planned for the other five convicted in the Bisbee Massacre. They were sentenced to hang. To string up five felons at one time was not commonplace, and in fact this hanging party was one of the most unusual in the West. Sheriff Jerome Ward even had printed invitations sent to special friends and officials, and picnics and other entertainment were planned, with the highlight to be the necktie party.

The five doomed men were Omer Sample, James "Tex" Howard, Dan Dowd, Daniel "Mick" Kelly, and William "Billy" Delaney. They were to swing early in the afternoon of March 28. Sheriff Ward had arranged to have about 500 guests in the jail yard, but many more spectators were expected in town for the event. To handle the extra crowd, bleachers were built outside the jail yard, so that this "second crowd" could watch the prisoners dangle.

This was indeed a holiday atmosphere. On March 27, for example, the world heavyweight champion John L. Sullivan was in Tombstone and he visited the condemned men. The men joked with the champion, asked if he could beat the Sheriff in a boxing match. Sullivan commented that Sheriff Ward "don't look like a fighter." Tex Howard replied to Sullivan, "Well, he ain't, but he'll knock five of us out in one round next Friday morning, just the same."[23]

Enter, now, Nellie Cashman. Reportedly she was very upset about the plans. Several of the "boys," Kelly and Delaney, were Irishmen, and Nellie felt it was wrong for the bleachers to be erected and have hundreds gawk and joke about the death and suffering of the Bisbee murderers. So, supposedly the night before the event she corralled a group of miners and they quietly tore down the bleachers, thus at least getting rid of some of the carnival atmosphere of the occasion.[24]

Even the briefest of sketches of Nellie's career includes the above scenario, but it is not based on any of the many accounts of the day. The correspondent of the Arizona *Daily Star* lent partial credibility to the story, as he quoted Delaney as saying:

> I am glad the miners tore down the platform outside of the jail yard that was to give so many people a free exhibition. Five hundred men sitting up there, in addition to eight hundred in the yard would make us a little shaky.[25]

However, the account in the San Diego *Sun* seems to have the more logical version. The "extra" balcony outside the jail yard had an admission price of a dollar and a half, "but the mob became indignant and tore it down. In the row which followed several persons were injured. One man had his leg broken, and another his arm. . . . With this exception, everything passed off quietly."[26] The men "swung," and following their deaths their bodies were delivered to the Tombstone priest Father Patrick Gallagher, as had been promised the doomed men.

The whole spectacle — the jail interviews, the joking, the grandstands, the last words from the gallows, the giving of the bodies to the priest — all was covered in great detail by the press of Tombstone, Tucson, and San Diego. Nellie's name never surfaced. She did not interview the prisoners, and there is no evidence that she was on hand for the hanging. We are left with the fact that the miners did indeed tear down the extra balcony-grandstand. Whether they did it out of compassion, or because they resented such a stiff admission price to witness a hanging, is unknown.

The only contemporary thread on which to hang this tale is in the reminiscences of John Pleasant Gray, who recorded that Nellie had considerable influence with the miners and even controlled their vote.[27] This may be an exaggeration, but Nellie could have passed the word to the miners that making a public spectacle of "good Irish boys" would not lead to any good.

The other event of 1884 that cloaks Nellie in a garb of compassion is said to have happened when she saved the life of E. B.

Gage, superintendent of the Grand Central Mining Company. Labor troubles in the mines and mills of Tombstone had led to threats of violence. Nellie heard that Gage's life was threatened, so she went to Gage's home at night, told him what was planned, and convinced him to leave town. She drove him to Benson, where he boarded the train for Tucson and safety. This scenario was reported in later years by two well-known Tombstone chroniclers, ex-mayor John Clum and Wells, Fargo undercover agent Fred Dodge.[28]

There was a serious labor disturbance in Tombstone early in August of 1884. Most members of the Miner's Union had refused to work unless conditions and wages were improved. The mining companies got along quite well without these men, hired non-union laborers, and kept working at full production. The union members began to grumble, then complain aloud, finally taking to the streets and were "exhibiting symptoms of a revengeful spirit." They threatened to burn and destroy if their situation was not ameliorated.[29]

An incident in Tombstone on August 8th triggered subsequent events. N. B. Lazard, walking towards the Head Center Mine, was yelled at by Union member Simon Washburn. After an exchange of words Lazard moved on, but Washburn shot at him — and missed — with his rifle. Lazard drew his pistol, returned the fire, and Washburn retreated. Sheriff Ward then took Lazard to jail for safe-keeping, but a mob collected. Finally the Sheriff and the mob came to an understanding: if any disturbance occurred, it would not take place in town.[30]

At 3:00 a.m. on August 9, a mob of forty to seventy men gathered, and headed towards the Grand Central. A foreman, C. W. Leach, spotted them near the hoisting works and watched as they approached the mine office. There a leader among the miners asked Leach if the mine owners would continue to pay $3 a day in wages. Leach answered by making a prisoner of the man, which infuriated the miners outside of the building. Shots rang out, and the miners withdrew some yards and peppered the building with as many as 100 rounds from pistols and rifles. At that point the Grand Central whistle was blown, which alarmed the entire town. The mob retreated.[31]

In the next few days tempers cooled somewhat, and the United States Army was asked to provide temporary law and order; a company of troops from the First Infantry Regiment from Fort Huachucha moved in to make certain there was no further trouble.[32] The only result for the Miner's Union was a bad image, as several of the Tombstone and Tucson newspapers condemned the men for the cowardly attack, acting "like thieves in the night."[33]

Where Nellie was during these hot times, or what she may or may not have done, cannot be gleaned from contemporary accounts. Furthermore, there is nothing in the accounts to indicate that Superintendent Gage was involved. The only credible link is a note in the *Arizona Daily Star* of August 16, stating that "E. B. Gage of the Grand Central has returned to Tombstone and will commence operations all along the line. He is accompanied by his wife."[34] Had he been chased out of Tombstone, saved by Nellie Cashman? If so, one wonders why his name was not mentioned in the many news accounts, or why he returned to the city-of-threats with his wife in tow.

Therefore, 1884 saw Nellie's reputation enhanced by showing compassion to men about to be hanged, and by saving a mining superintendent from certain death at the hands of a mob. But she seems to have done so only retroactively.

A curious circumstance is that in the many statements and interviews by and with Nellie Cashman, she never took credit for either event, never claimed to have been a participant in either show. How, then, does one account for such rewriting of history?

This may have been a "family conspiracy." Nellie's nephew Mike Cunningham became a prominent figure in Cochise County, affiliated with several banks in Bisbee and Douglas. His closest associate in Bisbee was another Irishman, William H. Brophy, an officer with the great Copper Queen Company in Bisbee. Nellie had introduced the two, and later they each married sisters of the

THE RUSS HOUSE!

MISS NELLIE CASHMAN, Manager.

Thoroughly Refitted. Everything New.
Eastern Waiters. White Cooks.

MEALS 25 CENTS AND UPWARDS.

Board $6.00 per Week, in Advance. $6.50, if Paid Monthly.

Goodbody family. Brophy's son, Frank Cullen Brophy, and Mike would both serve as presidents of the Bank of Douglas. In later years Cunningham fed Brophy much about Aunt Nellie, and Cunningham was also a prime source for John Clum in the late 1920s.[35]

Maybe Nellie had told Mike of these events; perhaps she was present; and perhaps Mike elaborated a bit over the years; and after Nellie's death the stories became part of Arizona history. The two events are important affairs in Arizona's past, but there is no contemporary evidence that Nellie Cashman was more than a bystander.

For much of 1884 and 1885 Nellie's tracks are hard to follow, as she seems to have left the Russ House. She was involved in a few purchases and sales of mining claims. In the middle of 1884 the Cashman sisters were involved in a complicated transaction. Kate Ward and Fanny Cunningham paid Nellie $3500 for a large assortment of furniture, food, clothing, and other materials at the Russ House. The detailed, three-page inventory includes such items as sugar, pillows, cots, coffee, jelly, peaches, brandy, sauce, and so forth. Even quantities were listed, as as "5 cases Sugar peas." This was Fanny's last gasp, as she tried once more to earn a living in Tombstone. Fanny quickly deteriorated, was taken to San Diego where she died on July 3, 1884, only weeks after the transaction. Nellie then continued to operate the Russ House.[36]

In February of 1885 she got involved in a drawn-out claims case, when John Smith [Smythe], a bartender in the Russ House, filed suit against her for back wages. Nellie filed a counter suit, claiming that Smith had neglected to pay his own room, board, and bar bills for months. Nellie was vindicated in the court action, but many months went by before both parties decided to end the legal snag.[37]

Near the end of 1885 it was announced that Nellie would again take charge of the Russ House. She had secured the services of an excellent cook, and "will have everything first-class as she always does."[38]

When the Russ House re-opened, she was listed as manager; meals were 25 cents, and upwards, board $6 a week, "everything new, thoroughly refitted." One other refinement could make idolaters of Nellie squirm; there were Eastern waiters, and "White Cooks."[39] This was bowing to the periodic pressure in Arizona communities to get rid of Chinese in any kind of employment.

Around this same time, in November of 1885, Nellie worked an interesting deal with E. B. Gage of the Grand Central Mine. She gave her rights to the Grand Central South and the Southern Bell claims for $600 in gold coin. The transaction also included the "Nellie Cashman Boarding House and Lodging House and all things appertaining to said House and Mines." She must have decided to make the Russ House serve as her living quarters.[40]

Less than two months after Nellie took over the Russ House, she was looking elsewhere again. Mrs. Pauline Jones of the International Restaurant and Nellie Cashman announced in early January that they would each move to Nogales and embark in the hotel and restaurant business. By January 10 the plans for both ladies had advanced. Mrs. Jones was to open the Palace Hotel and restaurant, with fifty-two rooms. Nellie Cashman rented a two-story building which had been a restaurant, which she planned to convert to a restaurant and lodging house.[41]

By early February Nellie was in business, and the Delmonico was operating in Nogales, with "clean and comfortable beds. . .

THE DELMONICO,

Nogales, A. T.

T. J. CUNNINGHAM..........Proprietor
MISS NELLIE CASHMAN.......Manager

This Hotel has been thoroughly renovated. The best table in the territory. Clean and comfortable beds.

Bar supplied with the finest Liquors and Cigars.

The Cunningham listed here was teenager Tom, Nellie's nephew. (Tombstone Epitaph, February 3, 1886)

the best table in the territory." And, knowing where the money was, she also boasted "the finest Liquors and Cigars." Although Nellie was listed as manager, the "Proprietor" was T. J. Cunningham. Here was Nellie, most likely following through on a death-bed promise to her sister Fanny. T. J. was Nellie's nephew Tom, born in 1872, and she was trying to launch him in a career at the ripe old age of fourteen or fifteen.[42]

Nogales had some potential at that time. Local newspapers wrote "great mining activity prevails," "building is still be actively prosecuted," "the Nogales Smelting Works is nearly completed."[43] Good silver and gold deposits had been found, and because Nogales was on the railroad between Tucson and all of

Sonora down to Hermosillo and Guaymas, the town had potential as a market center and as a location for a mining-milling-smelting complex. However, "potential" is not what Nellie found. The Nogales "boom" was less than that, and she quickly realized that there was no future for her there; no excitement was on the horizon in Nogales. A one-sentence entry in the Tombstone *Epitaph* of April 17, 1886, stated that "Nellie Cashman has sold her hotel at Nogales."

Discouraged with things in Arizona, Nellie went back to visit her mother in San Francisco. Within a matter of weeks she decided to give the territory one more swing, and she entered negotiations with Fred Maish, proprietor of the Palace Hotel in Tucson. An understanding was reached, and Nellie arrived in Tucson on June 28, to the welcome of the town: "The public may expect a boom now, as Miss Nellie understands how to run a hotel." She was not in charge of the hotel, only the dining rooms. This was billed as Tucson's most central, favorite stopping place, so Nellie was at least in the middle of some activity.[44]

Nellie's time in Tucson was to be short. As has been frequently demonstrated, Nellie had little geographical loyalty. Her allegiances were to excitement, to promise, to the optimism and loud noise of a thriving camp, town, or city. She was gradually concluding that such was no longer possible in Arizona Territory.

DELMONICO LODGINGS, Mrs T J Cunningham propr, Fremont bet 4th and 5th. First-class lodgings 50c

Dibble Henry, attorney at law, Fremont

DILLON & KENEALY, dry goods and men's furnishing goods, 212 Fifth

Diss & Richards, T & C Ice Co, cor Tough Nut and Fifth

Doran & Hearn, liquor saloon, 513 Allen

Drum Thomas J, attorney at law and court commissioner

Dunbar & Behan, livery stable, 318 Allen

Dunbar John O, county treasurer, Fremont

DUNN E C, physician and surgeon, 210 Fourth

EARLL, SMITH, CAMPBELL & ROBINSON, attorneys at law

Earll W, attorney at law, cor Fifth and Allen

Earp Virgil, city marshal, cor Fifth and Allen

Ekelund C P, saloon

Enright D, sign painter, 223 Fourth

ESCHMAN CLEMENTS, saloon and lunch house, 422 Allen

Farrington R E, groceries and provisions, 324 Fremont

FELTER A J, justice of the peace and attorney at law, 424 Fremont

Finney Geo W, house and sign painter, Fifth

FITCH THOMAS, attorney at law

FITZHENRY J C & CO, groceries and liquors, 216 Fifth

Fly C S, photographer and lodgings, 312 Fremont

Entries from McKenny's Pacific Coast Directory of 1882-83 with some Tombstone notables, including Virgil Earp, photographer C. S. Fly, attorney Tom Fitch, and the Dunbar livery stable, with Johnny Behan as partner. The top entry, Delmonico Lodgings, was bank-rolled by Nellie and was managed by her sister Fanny, Mrs. T. J. Cunningham.

V

NELLIE BURIES TOMBSTONE

FROM 1880 through 1886 Tombstone had been Nellie's headquarters in the Southwest, even though she had periodically explored and prospected as well as run hotels and restaurants in Tucson, Bisbee, Baja California, and Nogales. Most of that time she had been either owner, part owner, or manager of the Russ House, which provided her a consistent income and a place in the community. But too many factors were undermining Tombstone's future. A few serious accidents and fires in the mines and mills, the lessening richness of the ores, and a national policy which began to favor gold over silver were all bad news for Tombstone.

In early 1887, after a visit to her mother in San Francisco, Nellie set her sights on the younger mining community of Kingston, New Mexico. She stopped in Tombstone enroute and joined her friends Malcolm McQuade and Dennis McCarty in another mining deal; this time they located the Miners Dream claim in the Tombstone Mining District. She then went on to Kingston, a silver mining center in Sierra County in the southwest part of the state, discovered in 1882; a townsite had been surveyed in August of that year. Within a few years there were more than 2,000 people living in a raucous, fast-drinking mining town. A dance hall, casino, three hotels, and three newspapers, were constantly catalyzed by twenty-two saloons.[1]

Kingston was in the Black Range District, which was on the east slope of the Continental Divide, and the center of more than a dozen mines and mills. One aspect of the community that may have attracted Nellie Cashman was its Irishness, even though that could be said of most Western mining camps. Apparently quite a few of the men had come from the coal mines of Pennsylvania, where they had been affiliated with the Molly Maguires in some bitter labor disputes.

There were Irishmen galore in the Black Range District, where one claim was known as Erin Go Bragh Mining Company, and a monument had been set up on the claim "Tara Hall." One particularly tough group of characters was known as the "Four Neals," consisting of Neal McGarvey, Neal O'Gallegher, Neal McConway, and Neal O'Boyle.[2]

Nellie set up a hotel and boarding house in mid-1887, which was called the Cashman House. She was pretty much in the center of things, acting as a peacemaker for the rowdy crowd, and on occasion doing other duty. Once Neal O'Boyle was to go to a funeral as a pallbearer, but he was so bleary from alcohol that he needed a drink. According to the historian of the camp, McKenna, Nellie offered him a swig of Old Monk Spirits:

> 'Tis a tricky Yankee drink made of colored water and sugar. I feel that bad, Nellie, I don't think I can go to the funeral. Can't ye get me a dram of the real stuff, Nellie?
>
> "Come, Neal," said she, "I'll give you a drop from the sick folk's bottle," She led the way into the kitchen.
>
> "Nellie, ye have saved my life this day," exclaimed the grateful Neal, smacking his lips. "Shure, you're the fine coleen!"[3]

Another of the Irishmen in Kingston was Ed Doheney, later to become one of the world's richest men, involved in oil in the United States and Mexico, and accused in the famous Teapot Dome Scandal of the 1920s. The Nellie folklorists have Ed Doheney washing dishes at the Cashman House in Kingston, for change or tips.[4]

The reality was otherwise. Doheny was in his late twenties at the time, taught school, owned and worked several claims, and was one of the better informed mineral experts on the Black Range. He was the most reliable consultant for the hundreds of wandering and working Irishmen in Sierra County.[5]

As usual when Nellie was in a mining camp, she coordinated much of the work of the clergy, gathering funds, or heading up church bazaars and similar charities. McKenna points out that she was in charge of rounding up the Catholics when Father Eduardo came to town to say Mass.

If Kingston had excitement, it must have been before Nellie arrived. Before the end of 1888 she had again packed her bags and headed back to Tucson, to plan her next move. The New Mexico silver camp had not lived up to its promise.[6]

Nellie was probably visiting in Tombstone or Tucson when she heard and read about Arizona's next big mineral find. Around eighty miles due west of Phoenix, in the Harqua Hala Mountains, three prospectors hit a rich gold-in-quartz deposit. This was in the Centennial-Harrisburg vicinity, which was formerly in northern Yuma County, currently part of La Paz County.

The find was made in November of 1888, and by December all the territory's newspapers were heralding the discovery. Three men were involved in the strike — Harry Watton [or, Wattron], Bob Stein, and Mike Sullivan. They had apparently worked their way south from Mohave County and found the rich, black-looking ore which contained a considerable iron content as well as free gold. For years this area had been worked by Mexican placer miners, but they did not find the rich ore, which was a little deeper than their surface workings.

The excitement soon drew prospectors, miners, merchants, and a few sight-seekers from Tombstone, Tucson, Phoenix, and other Arizona communities. By late December there were about thirty people at the site, and merchants were loading up in Phoenix to supply the new camp. The two richest locations in the district were the Bonanza and the Golden Eagle claims.[7]

Nellie was ready to be in on the early, most profitable phase of this strike. Even though this was a gold-quartz rather than a gold-placer strike, and meant that large machinery would be required, an individual quick on the scene could profit. Good claims might be purchased, and a merchant with goods was well positioned to make a dollar. By late December of 1888, Nellie left Tombstone for Phoenix, where she loaded up on provisions and headed for the Harqua Hala Mountains.[8]

Huge chunks of rich gold ore were reported almost daily, and some idea of the hustle-bustle of the new camps can be gathered from these comments in the Phoenix *Daily Herald* of December 22, 1888, precisely when Nellie was enroute to Harqua Hala:

> A team loaded with passengers left yesterday, another team left today, tomorrow two leave and on Monday a four horse coach goes out. There is an immediate call for a stage line between this city and the new camp.

The same issue reported that Watton, Stein, and Sullivan were still hitting rich ore, getting $20,000 from two holes only seven feet deep. This is the kind of reporting that excites knowledgeable people.

Some of these "experts" included two former law enforcement officials from Pima County. Ex-Sheriff Charlie Shibell was one of the first on the scene, buying and working several claims. His ex-Deputy Sheriff, Wyatt Earp, living in San Diego at the time, bought alcohol, roulette tables, and other paraphernalia and headed for the Harqua Hala Mountains. Soon there were enough folks on hand to have a Tombstone reunion.[9]

The Phoenix *Daily Herald* of January 14, 1889, confirmed with tongue-in-cheek that Harqua Hala now was obviously providing "evidences of a big camp." They had their first shooting-killing, when Pete Barnes gunned Alonzo Johnson. The newspaper guessed the reason was "because there are five saloons on the ground and not a boarding house or any place to get grub in the camp."

With the hills starting to crawl with prospectors and miners, Nellie did what she could to answer their needs. She made at least three trips to Phoenix before mid-January to load up wagons with clothing, food, alcohol and other necessities.[10] According to one account "Miss Nellie Cashman will open a boarding house." She may have, but the majority of evidence regarding Nellie in the Harqua Hala rush suggests she spent most of her time working her claims and selling supplies.

An incident of late January suggests that Nellie was doing well in the new camp. On January 20, Nellie had gone to Phoenix and while attending Mass at the Catholic church realized she had dropped her pocket book. Asking Marshal Blankenship for help, they tried to retrace her steps, but Blankenship was not optimistic. Young twin boys, sons of A. Redeuill, on their way to Sunday School, found the pocket book loaded with currency and other personal papers. Their father heard about Nellie's loss and quickly saw to it that the pocket book was returned. Nellie, having "only a nickel left to her name," was delighted and gave each of the boys $10.[11] Four hundred dollars was a sizeable sum, and that of course did not include what she had in banks, the value of the claims she owned, or the supplies she had already purchased. Nellie Cashman was doing well in the Harqua Hala gold district.

And maybe she even fell in love. The following news item appeared in the Phoenix *Daily Herald* of February 23, 1889:

> Mike Sullivan, of the owners of the Bonanza mine at the Harqua Hala camp, left there yesterday in company with Miss Nellie Cashman, on their way to the nearest station where a minister could be obtained, in order to be made man and wife.

There was no elaboration on this statement by any of the newspapers, and in her many interviews Nellie never mentioned such an event, or a near-marriage. It was probably not a journalistic prank, as Sullivan, a wealthy, prominent miner was not to be laughed at, and any editor knew the pitfalls of making fun of

Nellie Cashman. In fact, about this time Sullivan had gone into Phoenix, where he was interviewed by the newspaper on the future of the mining camp, and Mr. Sullivan "is no enthusiastic but a cold matter of fact prospector and miner and his statements carry the evidence of facts."[12] About mines, not about marriages.

Nellie Cashman and Mike Sullivan could have been romantically involved. They were both Irish Catholics, both fascinated with and actively involved with mining, and both were among the leading characters in the new Harqua Hala mining camp. Whatever great attraction they had for each other must have quickly lapsed. Sullivan sold off his interests in July and headed back East, probably never to return.[13]

In early March, shortly after the non-marriage, Nellie Cashman wrote a mining report which was published in the *Arizona Daily Star*. The letter, in addition to being a sound summary of the origins of the camp, also indicates Nellie's good command of mining geology, and her attitude of what is important in a thriving mining community. The fact that such a detailed analysis would be published under her name suggests the high credibility she had in the mining culture of the Southwest:

> On arrival here from the rich mines of Messrs. Sullivan, Stein and Watron, situated in the Granite-wash range, a few miles south of the Harqua Hala, being my third trip, you would like to know what change has been made in developments since I first saw the prospects. As is usual from the number of years I have spent visiting almost all of the mining excitements of this coast, "staying with them that have proved good," I found an unlimited number of old friends, who strove among themselves in the heartiest manner, greeting and entertaining me to their best facilities, making my visits among them a source of pleasure to me which is ever treasured.
>
> When I first saw the three places where the three strikes were made there was but little work done, even that was not necessary to convince me from the rich croppings that they would ultimately prove to be mines of great extent

and richness. The period entervening from my first to second visit, which was not long, I found comparatively no work done. My reasons why, I suppose, was from the great number of visitors who came in the excitement took all the owners' time to show and guard their properties, which are separated one and a half miles apart. From what I have seen after four days stay visiting most of the principal mines and locating in the meantime, I will endeavor to state here as accurately as my judgement will admit.

Messrs. Sullivan and Co. known as the principal locators, have three gold mines No. 1, is not being worked at present. No. 2, is being worked, cross-cutting and sinking inclines on supposed foot-wall; when 20 feet is sunk will cross-cut to find hanging wall. This mine has held its own, having rock sacked that goes $1800 per ton. I think when this mine has obtained its proposed development it will be a finer mine than Tombstone ever had.

In proximity to this mine (side line boundary) is an immense rich carbonate mine, no work being done at present. The ore assays $49.50 in gold and $34.75 in silver and 60 per cent lead per ton. One and a half miles from this mine, on I think, the same range is the mine where this sample I give you was produced and as well as the many rich nuggets were found on the surface.

No. 3 is being worked sinking incline down 20 feet, I am told, all in ore of the richest kind. I don't know how wide this body of ore is, it seems to be the whole of one side of the hill on which it is situated. I would like to have the gold that the dump contains, which I am sure would be over $30,000. They intend sinking 50 or 60 feet and see by drifting if there are any walls, for this is the finest and largest ledge of exceedingly rich gold ore I have ever seen and if it should hold out to proposed depth this mine alone will make as good a mining town as Tombstone or Bodie ever were. Between these mines I have an interest in one on the same range and trend which I intend working soon; for Mr. John Quinn & Co. are developing a promising mine next to mine. The Messrs. Quinn have other very fine grounds in many places which they are developing now. Messrs.

> Cochrane & Co. have obtained from the first extension of No. 1 and No. 2, to the north, exceeding fine ores in developments.
>
> They likewise have a number of choice locations which they are working, and showing up. Messrs. Halpin, Owen and partners have many good locations. James Coil [Coyle], (Ex-Marshal of Tombstone) also has many good locations which will invariably become of great value soon, I think. For capitalists will never let such chances long remain uninvested. Such are the views of one who has truthfully given her honest opinion publically stated, your friend.
>
> NELLIE CASHMAN[14]

One of the bonuses in this interesting account is that the reader gets a dose of Nellie's philosophy of mining camps and does not have to rely on the historian or biographer. She believed in "staying with them that have proved good."

The Harqua Hala mountains would provide good gold and some rich returns for the next two decades, but neither Nellie, the original discoverers, nor those early in the camp would figure in those developments. After all, the Harqua Hala was, basically, a gold-in-quartz camp. Once the best claims were obtained, the small operator had no place in the district. The only men needed were a hundred or so laborers for the mines and mills.

Watton, Stein, and Sullivan all sold out in 1889, partially to a San Francisco syndicate. However, the complexity of the ores demanded huge expenditures and extravagant machinery. Aside from a stamp mill erected in 1891, nothing important was done in the Harqua Hala District until June of 1893, when some London investors got interested and formed the Harquahala Gold Mining Co., Ltd. English money would remodel the mills, bring in better machinery, sink deeper shafts, and install a cyanide plant.[15]

By late April of 1889, Nellie Cashman was in Tucson, where she had been staying for several weeks recovering from an illness.[16] She talked about returning to Harqua Hala, but she may not have done so. She sold her claims and may have done well in doing so. Her best property was adjacent to the claims of John

Quinn and his brothers, and for the next few years the Quinns did as well as anyone in the district.

Some comment about the Quinns and their claims could be useful in understanding Nellie, her claims, and her success. The Quinn brothers were early on the scene; one report has them establishing their claims the day after the fabulous find by Wattron, Stein, and Sullivan. Another report has their claims "at either end of the extensions" of the original founders of the Harqua Hala strike. A further statement, during the height of the rush, leaves no question about the quality of their ground: "The Quinn Boys have some of the best claims in camp."[17] Nellie Cashman's claims, then, were adjacent to those of the original strike, and also next to those of the Quinn Brothers, among the richest in the region.

By summer the Harqua Hala was still in the news, but only because of financing, and hopes for the future. The camp's "boom" was over. Charlie Shibell returned to Tucson, and Wyatt Earp, frustrated at his inability to get water and seeing that the new camp had stopped growing, headed back to San Diego. This was not the type of place to keep Nellie Cashman's interest.

There is very little news of Nellie in the period after she left Harqua Hala in late spring of 1889 until more specifics surface in 1894. However, one news item from late 1889 is intriguing as well as puzzling. The *Arizona Daily Star* of November 23 stated:

> Miss Nellie Cashman is visiting with Mrs. E. J. Smith of this city. She has just returned from a trip to Africa and is here in search of a crew to accompany her to explore a hitherto unheard of diamond region. Reasons for Arizona: men here are more reliable, courageous, and can endure more hardships than any she has met in her travels.[18]

Did she go to South Africa? If so, it is difficult to see when she found the time. In March, Nellie Cashman "of Pima County" was involved in the sale of several mining claims, for which she received $1200. For much of April she was bedridden in Tucson.

In May and July Nellie, acting as administrator for the estate of her nephews Tom and Mike Cunningham, sold a few claims for them, including the Miners Dream near Tombstone. Her trail is dim for most of July through November, so the scenario of a trip to the South African diamond fields was possible, and she certainly had the money from her Harqua Hala success. But no further information is available on this chapter of her life.[19]

Almost as soon as she arrived in Tucson, she departed for the Northwest, visiting Oregon and Washington. She was back in Tucson by December 6, where it was reported she would probably open a boarding house in the Harqua Hala District.[20] Apparently she did so, as there are several mentions of the fact in a Tombstone newspaper.[21] But with gypsy-like moves she turned up in Prescott on January 23, 1890, supposedly with the intention of opening a hotel or a restaurant.[22] There was considerable mining excitement in greater Prescott at this time, due to new finds and to efficient railroad transportation, so Nellie may have been temporarily attracted. But there is no evidence that she remained.[23]

Where she went, what she did in the next few years, is uncertain. One attempt at completing a chronology of these years suggests that she tried Bodie, California, and the Coeur d'Alene district of Idaho, and perhaps also Jerome, Arizona.[24]

There are other hints, though, in the statements of Mike Cunningham. Nellie was a surrogate mother to the Cunningham children, and these years, when they were in their teens, caused her to devote more time to them. Mike recalled prospecting, mining, and living in wooded areas of Idaho, Wyoming, and Montana during this era, sometimes in a mountain cabin "where snow was piled nine feet high about the house." Once in Montana, Mike recalled, he left her "because I just couldn't keep up with her. We were always on the move, looking for gold and silver."[25]

By 1894 Nellie's trail can be picked up with more detail, but her pace did not lessen. There was a frantic energy in her moves, now, as she was looking for something steady, but not finding it. In the late summer of 1894 Nellie opened a restaurant and boarding house in Jerome, Yavapai County, Arizona Territory. Jerome was an important copper production center to the north and east of Prescott. This was a mistake and quickly remedied, as by December she had closed her operation there.[26]

By the end of January Nellie had moved her hopes to Prescott, where she took over the Elite Restaurant "with a healthy patronage which is bound to increase."[27] Less than two months later the Elite was closed, and Nellie was obviously getting disgusted with the lack of excitement in Arizona Territory.

On the spur of the moment she headed for the mining district of Juneau, Alaska, not to mine, nor go into business, just to visit. She stayed only a few days but was fortunate to meet acquaintances from her days in the nearby Cassiar District of British Columbia.

> You see many of the boys were just about to go into the interior to prospect, and when I arrived they appeared glad to see me. They built a huge log fire and made me sit by it and talk of the old Cassiar days. They said it looked natural to see me sitting by a camp fire as in the days of yore.[28]

She arrived back in Phoenix in late May of 1895, but the comradeship of the North would remain with her.

Not one to tarry, Nellie got a reading on what was the most active camp around, and by June she was on her way to Minas Prietas, Sonora. She had intended to open a boarding house in that gold camp, which was thirty miles to the east and south of Hermosillo. She did go there, but almost immediately returned to Arizona.

The Minas Prietas operation was huge, but things had been pretty well established by time Nellie examined the place. Two huge ore bodies had been developed by the Creston-Colorado and the Grand Central companies, with all the latest in mining equipment and milling methods. A New York syndicate headed by John Gates operated most of the Minas Prietas works. Other Tombstone personalities were also on the scene when Nellie arrived, and Dr. George Goodfellow left his practice in Tombstone and Nogales to practice there for a few months.[29]

She went up to Cripple Creek, Colorado, for a few weeks, just like the Earp brothers and several thousand other "boomers," but she quickly realized she was too late for the Cripple Creek rush.[30] She stopped for a visit in Prescott in early October of 1895, visited Casa Grande south of Phoenix, then arrived in Tucson on October 24 on horseback, which impressed the editor of the *Arizona Daily Star*. She was praised for her talent, her years of experience in the mines, and was "always among the foremost on the ground and saw the boom set in."[31] The story was picked up by the Los Angeles *Times*, which added:

> When in camp, and she has been continuously in camp, she has lived a miner's life and has always enjoyed the deepest respect of the men with whom she has been associated.[32]

A copper camp called next, and Nellie headed for Globe, Arizona Territory, which was emerging as a leading mining, milling, and smelting town. By late October of 1895 she was ready for business, having visited some of the Cunningham relatives in Tucson and Bisbee first. She worked an arrangement with Lewis Robinson who had a property on Broad Street, Globe, and opened the Buffalo Hotel and Restaurant in February of 1896. The notice in the Globe *Arizona Silverbelt* of February 2, 1896, promised "table of excellent quality . . . lunches a specialty."[33]

Although off to a good start, the local economy practically came to a halt in June of that year, when the Old Dominion Mine shut down. Most of the miners and millmen were out of work, and a Tucson editor bemoaned the fact that Globe's streets were "filled with idle men."[34]

Her next move was to Yuma, Arizona Territory, the seat of Yuma County, on the Colorado River. Yuma was an important transit point; east-west coaches had been important, and now the Southern Pacific was an important link with the Pacific Coast.

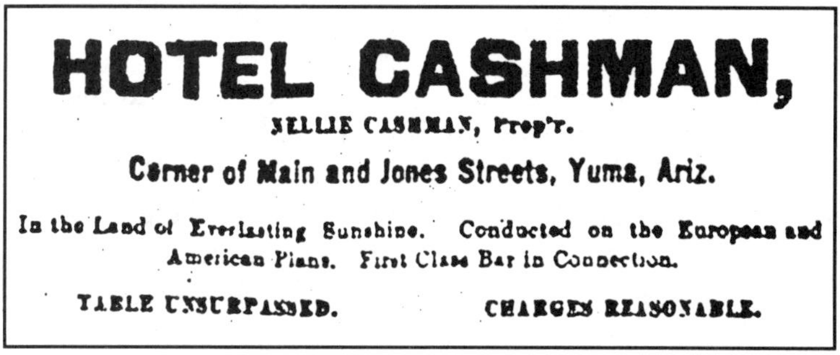

(Arizona Sentinel, July 24, 1897)

Steamer traffic on the Colorado River was periodically heavy, as mining machinery, ore for processing, and supplies and passengers used this artery frequently. The Territorial Prison was in Yuma, and in the vicinity there were more than forty working or promising copper, gold, mercury, and silver mines.

Nellie arrived in Yuma in early July of 1897 and negotiated for control of the Jones Hotel.[35] She opened for business in a few weeks, and named the establishment Hotel Cashman, meals on the European and American plan, table unsurpassed, and "first-class bar in connection."[36] The hotel was on the corner of Main and Jones Streets, and Nellie was ready for business. She was billed in the local newspaper as "no novice," but one who had conducted similar restaurants and hotels "in every mining camp in the west . . . will aim to give the public first class service in every particular."[37]

Living with Nellie in Yuma was a niece, Mamie Cunningham, who may also have worked in the Hotel Cashman. However, Mamie's mind was on other matters, which led to a happy occasion in early November. She had fallen in love with Miles Archibald, a guard at Yuma Territorial Prison. The young couple were married in Los Angeles and returned to make their home in Yuma. The wedding may have eased Nellie's situation considerably, because she had much to do with the raising of her sister's children. Within weeks after the wedding, advertisements for the Hotel Cashman no longer appeared in the local newspaper.[38]

The gypsy-like wanderings for Nellie were about to end, but neither because of the wonders of the land of sunshine in Yuma, nor because of the increasing patronage there of the Hotel Cashman. In the far north, in places like the Yukon and the Klondike, names unknown to most of humanity at that time, magnificent discoveries of gold startled the world, and an avid newspaper reader in Yuma would become ensnared in one of the world's most memorable gold rushes.

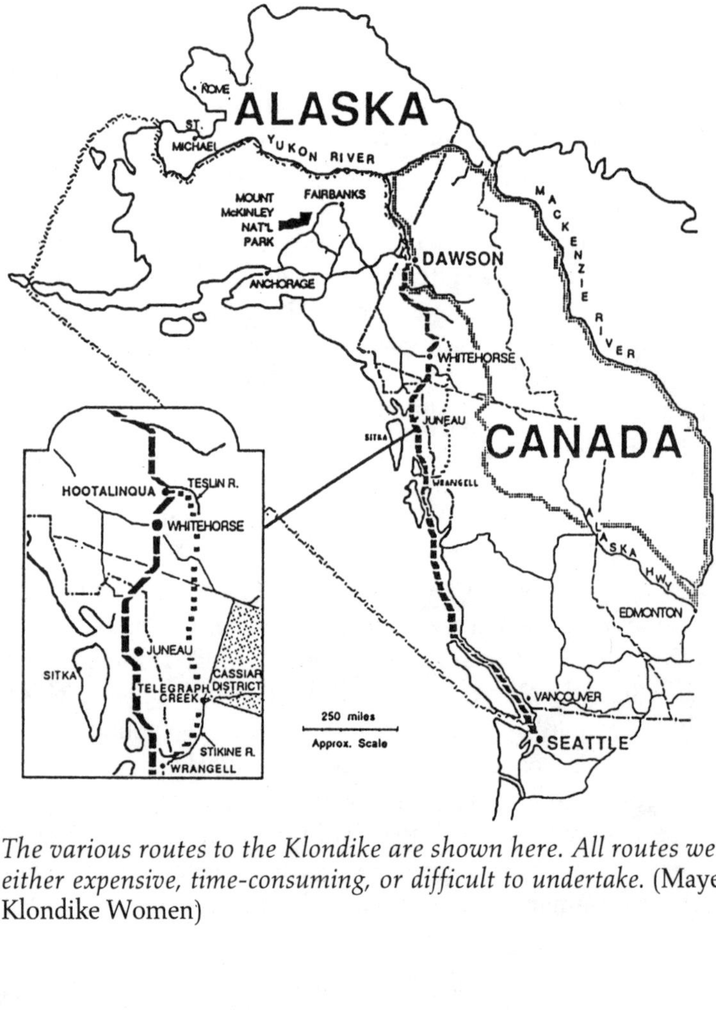

The various routes to the Klondike are shown here. All routes were either expensive, time-consuming, or difficult to undertake. (Mayer, Klondike Women)

VI

FROM DESERT SAND TO GLACIER PEAKS

MUCH of southern Arizona where Nellie Cashman spent twenty years is desert. Extreme summer heat, hundreds of varieties of cacti, and stretches of land for miles without evidence of tree or bush characterize much of the land. The last twenty-five years of Nellie's life was spent at or above the Arctic Circle, and she usually wintered there also. Snow drifts, temperatures usually below zero, and winters that seemed to last most of the year were the norm for the Yukon and northern Alaska. Nellie Cashman adjusted easily to both these disparate climes, and for reasons which will become evident came to favor the land of ice and cold over the desert.

Gold was known to occur in the north, and productive mines had been worked in British Columbia and Alaska. By the mid-1890s the gold mines and mills of Juneau were major producers. Exploration and prospecting had gone on in the interior, but with only modest success. Even the discoverer of Tombstone, Ed Schieffelin, believed in the northern fields and led an expedition up the Yukon River in 1882-83. His party found some gold, but Scheiffelin felt the distance from major population centers, hazards of transportation, and the horrible climate meant that central Alaska would never be a major producer of gold.[1]

Over the years fur traders, government agents, United States military personnel, and the hardy occasional wanderer looked for precious metals. Employees of the Alaska Commercial Company had the most excuses for being in the interior, and they kept in contact with Eskimo and Indian communities, trying to find signs of gold and silver.

By the mid-1890s, there were a number of working gold camps on the middle and upper Yukon River in Alaska. Rampart was a settlement on the Yukon, about half way from the Pacific Ocean to the Canadian border, and farther up the river were Circle City and Fortymile Post, each with about 300 inhabitants. Quite a bit of prospecting and gold placer mining took place between Rampart and the border of Yukon Territory. Geologist Robert Stein wrote a summary of these gold regions based on travel in 1894-95, and referred to the area as "the Yukon gold fields," but he was referring to territory touched by the river in United States jurisdiction in Alaska, not in the Canadian Yukon. Also, he was not optimistic about the future of the placers; "for every dollar obtained, $2 had to be spent."[2]

The big discovery came in 1896, not in Alaska, but a bit farther to the east, in Canada's Yukon District. George Carmack, his Indian wife Kate, and her relatives Skookum Jim and Tagish Charley were hunting, fishing, and prospecting around the headwaters of the Yukon. Two years earlier, at the trading post of Ogilvie, on Sixty Mile River, merchant Joe Ladue had grubstaked several prospectors and miners, including one Robert Henderson. In the summer of 1896 Henderson found gold in the Yukon, told Carmack and his party the general location, then returned to Ogilvie and told Ladue about the discovery. Ladue spread the word, and soon the outside world knew of the Yukon's riches.[3]

Meanwhile, Carmack got busy, and on Bonanza Creek he found fantastic riches on August 17, 1896. He went west to the Canadian Barracks at Forty Mile, and on August 20, 1896, the first claims were filed. The Discovery Claim belonged to Carmack, and he also filed for No. 1, below; Skookum Jim received No. 1, above; and Tagish Charley got No. 2, below.[4]

The history of the Yukon is underlain with bitterness, as

Henderson and Carmack disputed each other's words, and apparently Henderson bristled because Carmack rewarded his wife's Indian relatives with major claims. No matter, the ground was rich, the word was out, and the public clamored for news of this northern bonanza. Joe Ladue lost no time, and rafted his sawmill and a load of lumber to the junction of the Klondike and the Yukon, where he built a warehouse and cabin that served as home and saloon. Because he was early on the scene, he obtained the best land. He set up a townsite and named the place "DAWSON" after the well known Canadian geologist.[5]

After the discovery and claim filing of August 1896, some placer mining was done in the winter and spring, but not many people were present. Soon hardy prospectors, hearing the news in other parts of the north, found their way to Bonanza and Klondike Rivers and began work. The first official word to the "outside" came on June 5, 1897, in a report to the Canadian government by William Ogilvie, who was surveying the Canadian-United States border.

Seattle was shocked on July 16 when the "Portland" docked, and some of the miners on board waved bags of gold dust and nuggets at the people gathered at dockside. Soon a regular stream of ships entering Seattle and San Francisco carried returning miners from the Yukon country, many of them new millionaires.

Practically every newspaper in the country carried daily or weekly stories about the new bonanza. For example, the Los Angeles *Times* throughout the summer not only carried Klondike-Yukon stories, but also reported on the impact of the new gold strike in places like Prescott, Phoenix, Santa Fe, Albuquerque, and other southwestern cities. This was exciting news, especially in mining country.[6]

Such news and talk about the north country started Nellie planning at once. Her local newspaper, the Yuma *Arizona Sentinel*, kept its readers informed on the Yukon strike. In the issue of August 21, 1897, for example, there were several stories about the Klondike, and the lead editorial was on "THE NEXT EXCITEMENT," trying to understand the Yuma boys who seemed willing to head for the Arctic harshness. One of the first notices of Nellie's intentions was published in the *Arizona Daily Citizen* of September 15, 1897, stating that Nellie Cashman was talking of organizing a company to go to the Klondike. For the next several months she discussed the project with a few friends, talked to potential backers, and decided to leave the Yuma desert behind her to head for the Arctic placer fields. The following news note from Tucson appeared in most of the Arizona newspapers in November:

> Miss Nellie Cashman, one of the most favorably known women in Arizona, arrived from Yuma yesterday. Miss Nellie is preparing to organize a company for gold mining in Alaska, where she has visited three times. Her many friends in Arizona will wish her success, for during her twenty years residence in the Territory she has made several fortunes, all of which have gone for charity.[7]

Nellie couldn't organize "a party," but several sources later stated that she was grubstaked for the Yukon by James Finley of Tucson. A report that her young nephew Mike Cunningham of Bisbee sponsored her trip is most unlikely. By February of 1898, Nellie Cashman was ready, and she left Tucson and headed for San Francisco, then went on to Victoria, where she would decide which route to take to the gold fields.[8]

In many of the mining camps in which Nellie lived she was the only woman, or one of a few; the others were often soiled doves. In the Cassiar gold fields of British Columbia, and in Arizona's Harqua Hala, or in the gold placers of Baja California Nellie represented the only presence of decent femininity.

Nellie was not to be in the forefront of the female pioneers to the Klondike. At least five other women vied for the title of "first woman on the Klondike." The first was Kate Mason Carmack, the Tagish Indian wife of George Carmack. Other early pioneers

Dawson was an important port. These steamers were for travel on the Yukon River, down to the Pacific, while smaller craft were for going up the Klondike, Bonanza, and other tributaries. (Archives, Sisters of St. Ann, Victoria)

included Mrs. Thomas Lippy, whose husband mined and built the first cabin on El Dorado Creek, next to the Klondike River; Salome Lippy returned to California a wealthy woman in August of 1897, half a year before Nellie arrived up north. Mrs. Thomas Berry was another early pioneer, and journalist Emma Kelly reached Dawson in November of 1897. Belinda Mulrooney, with hotel and restaurant experience in San Francisco, was a stewardess on a northern steamship line, heard early of the gold strike, and before the end of 1896 was on the ground, filing claims and opening a restaurant.[9]

Dawson and its environs would have almost 40,000 people by 1904, and a few thousand of them were women. In fact, over a thousand women were actively involved in mining, either in the business end, or in the actual working of the placers. Nellie's gender would scarcely be noticed in such an environment. However, Nellie's other characteristics would continue to set her apart.

Nellie's plans to organize an expedition of prospectors, paid for by Arizona backers, did not materialize. She set off alone for San Francisco, where she was to meet with her nephew Tom Cunningham and another young man. They did not meet there, so Nellie sailed on to Victoria, British Columbia, arriving on February 14.

Seattle became the early shipping center for the Yukon, but Nellie preferred the city she knew well, Victoria. Her arrival was heralded. The *British Daily Colonist* published a lengthy article on her, headlined "Pioneer Woman Miner." Nellie's years in the 1870s at Wrangell and in the Cassiar District were praised, for her dauntless energy, and for her compassion. She was described as "a lithe, active looking woman, with jet black hair, and possesses all the vivacity and enthusiasm of a young girl."

She was staying at the Burns House and specified that she had stopped in Victoria because she had fond memories of the place, where in the past she had outfitted with the local merchants and had received full satisfaction. Her plans were to meet up with her nephew Tom and possibly a few other companions, head north through the Stikine River route, and on to Dawson. She intended to prospect, to mine, "but I will probably vary that occupation by engaging in business."[10]

While in Victoria Nellie met with many old friends from her Cassiar days, and one of them provided her with a written personal testimonial. This was the Honorable Theodore Davie, Chief Justice of British Columbia. He knew her during the Cassiar era and witnessed her efforts to help the sick, starving community. A portion of his letter read:

> I was much pleased to meet you again here in Victoria and have no doubt that the unflinching courage and determination which have been yours in the past years, will likewise guide you to success and fortune in the perilous trip to the Yukon, which, in mid-winter, you are about to take.[11]

While Nellie awaited her nephew Tom Cunningham in Victoria, Tom Cunningham showed up, but it was not the Tom she expected. Nellie had sent a letter to a Seattle merchant with $10, asking him to keep alert for Tom Cunningham. Through the local postoffice he located Tom Cunningham, gave him Nellie's letter and $10, and Tom took a steamer to Victoria.

In Victoria the bell-hop told Nellie that her nephew had arrived. "You are Miss Cashman; I am your nephew." Nellie stared at him; the young man became flustered, and Nellie stated "Your name is all right, but you don't look like my nephew ought to." The impersonator quickly caved in, said he tried to cash in on the fact that his name was also Tom Cunningham, and offered to return the $1 he had remaining out of the $10. Nellie could have had the boy arrested for impersonation, but let him off and told him to keep the $1. Nellie did not suspect foul play, and still expected her nephew Tom to meet her at Wrangell.[12]

There was no easy, short way to the Yukon. Of the four routes, the overland from Edmonton was the longest, most tiring, and was often deadly; even Canadians preferred other routes. The water trip was longest, easiest, and unreliable. This meant shipping from Seattle up to the Yukon river, then on smaller craft going more than a thousand miles to the headwaters, but one could not count on open river because of ice and floods. This water route was slightly more than 4,000 miles.

There were two overland routes. One was to begin at Wrangell, go up the Stikine River, through the Cassiar District, then overland by trails to Dawson, a distance of 850 miles. The other was to head for Skagway, then climb over two of the most famous, steep passes in North America, across a series of rivers and lakes, finally arriving in Dawson, a challenging trip of more than 600 miles.

Nellie, while in Victoria, had selected the Stikine River route. She was very familiar with the terrain and was even looking forward to the trip. She left Victoria on the "S.S. Centennial" on March 9 and arrived at Wrangell on March 13. Here she had two disappointments. Her nephew Tom was not there, and the news of the Stikine River route was bad. Mild temperatures had turned the melting snows into masses of slush, making walking with supplies for hundreds of miles practically impossible. She took the "Centennial" up to Skagway, arriving there around March 20. Nephew Tom was nowhere to be seen, so Nellie planned at once to tackle the Chilkoot Pass and head into the Yukon Territory.[13]

The trek up the Chilkoot Pass is one of the best known endurance hikes in human history. The photographs of the long, slender line of bodies slowly making their way up the steep, snow covered trail still make us ponder the strength of man, as well as the lure of gold. More than 100,000 people arrived at the foot of Chilkoot Pass; only about half had the stamina to continue. Very few of these hardy folks were female, and practically none of the females were Nellie's age, fifty-four.

Nellie Cashman knew she would make it, so laden with 900 pounds of provisions she headed up the pass. On a good day a traveler could make this hike in six to eight hours. But there were no good days, because each individual had to have at least 900 pounds of provisions, which meant arranging them in a variety of packs and containers, and returning to various points along the trail and gradually moving the packages farther up the trail.

At the top of the Pass were units of the Northwest Mounted Police, who inspected the provisions and verified that each traveler had at least 900 pounds, to enable him to survive at least a year. One report suggested Nellie did not have the required amount of provisions, but talked the Mounted Police into letting her proceed; "she had the gift of palaver and was a most convincing talker."[14] Some people traveled in groups, so were able to make good time; perhaps the entire Pass could be negotiated in five to ten days. People traveling alone needed two to three

Getting ready to scale Chilkoot Pass, one of the most famous and difficult treks in mining history. (Alaska Historical Society) ▶

weeks to climb and make countless return trips for their provisions. Nellie did most of her hauling, but from time to time was given a hand. She made the tortuous trip in less than three weeks, reached the top, got approval, then advanced as far as Lake Lindeman, Lake Bennett, Whitehorse Rapids, and Lake Laberge, where all travel was stopped for a week or so.

Near Lake Laberge several travelers overtook a prospector struggling with a heavily loaded sled pulled by a lone dog; it was Nellie Cashman. The travelers allowed her to hook on to their sled for the next few miles. Nellie was wearing a heavy Mackinaw coat, trousers, heavy boots, and fur cap. Although they admired her spunk, she was not permitted to camp with the men "on account of her sex," and after eating with them she moved on alone. The alternatingly kind and unkind behavior of the men was not based on like and dislike of Nellie. Woods, one of the group, described her as a "lone, undaunted figure in the wilderness, who had surmounted all the obstacles with which nature had beset her path," and he spoke highly of her when they met again in Dawson. Yet, a lone, aging female in the rugged northern mountains was an odd sight and caused many travelers to be confused.[15]

When the ice broke in the Yukon River, Nellie and the others traveled on barges down the mighty river, over dangerous rapids and finally arrived in Dawson in mid-April of 1898.[16] A year after she first heard of the Klondike gold strike, Nellie Cashman was on the ground, ready to become all taken up in another frenzy of mining camp living.

The gold mineralization in the Klondike probably dated from the middle Mesozoic Age, about 100 million years ago. The break-up of the host quartz and other rocks took place through the erosion caused by wind, rain, ice, and snow working over the ages. Geologists believe that most of the gold was of local origin, meaning that even though it was placer or alluvial gold, it had not moved far from its point of origin, but was the product of severe deterioration of rock over time. One of the first in-depth studies of the district was by J. B. Tyrrell of the Canadian Geological Survey, who concluded:

> The Klondike district owes its phenomenally rich placers not to the wearing down of highly mineralized gold-bearing veins or lodes, but rather to the favorable conditions of long continued and uninterrupted concentration from a great mass of rock that contained only very minute quantities of gold.[17]

Much of the gold had been and was constantly being moved by water, which accounted for the large amount of stream gravels containing gold. Yet, there was also terrace gold, referred to as bench deposits. These rich deposits were on higher ground, remnants of streams long disappeared, of uplifts and other land shifting. Gold in the Yukon, in other words, could be found in a variety of locations and elevations.

Gold was first located by sight or by panning, then further work was done by moving earth with water — the hydraulic methods — or by drifting, that is, sinking a shaft to bedrock, then following the rich gravels and sands by going or drifting in their paths. Almost from the beginning in the Yukon, miners took advantage of the frozen ground. They built fires underground, which thawed the drifts, which could then be hacked at, removing soil to be worked on the surface. Timbering was not common, as the frozen ground meant that safety was assured by nature. In a matter of months miners evolved a more practical, efficient method. Steam heat created from boilers was used. The steam would be forced through pipes, and the thaw point, aimed at the walls of the drift, softened the ground so that it could be worked.

The system of prospecting, discovery, and filing claims was based on an "above" and "below" concept. On a typical river, the point of first discovery, became Discovery. Then, advancing above Discovery, the numbering was No. 1 Above, No. 2 Above, and so forth. If a miner were on Bonanza Creek, working No. 17 Below,

Dawson in 1899, the maturing gold camp. (Archives, Sisters of St. Ann, Victoria)

he was seventeen claims below that of the original discovery. On Bonanza Creek the claims averaged ten claims to the mile.

Klondike miners found it much easier to work in the winter. Transport with dogs was much easier, and there was no worry of flood or other water damage. Because of the frozen ground, there was little difference when sinking shafts and working drifts in winter or summer. The Dawson area, in other words, would never be deserted in winter.

Of the tens of thousands of stampeders who flocked to the Klondike by the four different routes, at least 20,000 were on hand by early 1899; 5,000 were in Dawson proper or nearby, and about 1,000 of these were women. Almost from phase one, Dawson was to have some of the accoutrements of civilization: saloons, schools, newspapers, churches, and hospitals.[18]

One of the chief cultural and social pioneers in the area was the Jesuit priest, Fr. William H. Judge. He had served in the Alaskan gold districts of Rampart and Circle City, and with the rush to the Klondike knew that facilities would be needed there. He arrived in the early phase, erected a church and St. Mary's Hospital, and convinced the Sisters of St. Ann of Victoria to participate in the operation of the hospital; they arrived in June of 1898. A church and hospital fire late in 1898 weakened Fr. Judge, and he died in Dawson on January 16, 1899, victim of a flu epidemic in the community.[19]

Nellie arrived in Dawson in late April of 1898, and it took weeks, probably months before she could investigate and plan her moves. However, one thing was obvious from the beginning. Nellie had apparently fallen under the spell of Father Judge and the newly arrived physician, Dr. W. T. Barrett. The Yukon *Midnight Sun* of June 18, 1898, carried the following news note:

> Misses Nellie Cashman and Georgie L. Osborne returned Friday morning from a trip up the creek, where they solicited for the benefit of the hospital. They were quite successful and speak in grateful terms of the generosity and liberal donations of the miners. Miss Cashman is the pioneer woman in this country and is widely known for her good deeds.

At this time St. Mary's was the only hospital in town, hopelessly inadequate to serve such an expanding population; Good Samaritan Hospital would open before the end of the year. Nellie's good work was important in financing the three-storey addition to the hospital in 1898. She was further heartened to learn that the Sisters of St. Ann were to arrive and take over; she had helped this religious order build their first hospital in Victoria in 1875.[20]

Nellie's charity-prowling was not limited to the outlying districts. A visitor to Dawson in June of 1898 was using a photo studio's darkroom, when he heard in the outer office "the distinct tones of that rich Irish brogue." It was Nellie Cashman, twist-

For a year or so Nellie operated a restaurant and small grocery next to the Hotel Donovan, shown on the right. (Archives, Sisters of St. Ann, Victoria)

ing the arm of the photographer to donate to the hospital. The visitor was John Clum, whom Nellie greeted with "Hello Mayor Clum!" Clum, ex-mayor of Tombstone, was a U.S. Postal Inspector in Alaska and had no business in Dawson, other than being attracted to the goings-on in this boom town. Clum described Nellie as "robust, active, prosperous and popular."[21]

For the next six years Nellie was engaged simultaneously in mining and the restaurant business, and according to correspondence, court records, and newspaper accounts, most of her time was spent in her placer claims. The records tend to be confusing, because as usual, if something didn't work out, Nellie dropped it and started over. She ran at least five different restaurants while in Dawson. She claims her first was called the Delmonico, a short order place, meals anywhere from $2 to $6. Yet, the Dawson newspaper in late 1898 reported that "Miss Nellie Cashman, proprietress," had just opened the Cassiar Restaurant. When Clum saw her in June of 1898 she was running a restaurant adjacent to the Donovan Hotel. Two years later she was listed as having just moved her restaurant from Second Avenue and Second Street to the Donovan Hotel. In 1902-03 she was living out of town on Bonanza Creek, also operating a restaurant.[22]

Her nephew, Tom Cunningham, did catch up with Aunt Nellie in Dawson in the summer of 1898. They would work together for a few years, as Tom helped with the claims as well as the restaurants. By late September of 1898, a few months after arriving on the Klondike, Nellie was about to open her first restaurant. She was the owner of and worker of four mining claims and already had "a tomato can loaded with dust."[23]

Nellie was to encounter many pitfalls and problems in the Yukon. In the long run she profited, but this was not a mining camp in which she had her own way, or one in which her reputation remained unscathed.

A month or so after she arrived, she, her nephew Tom, and a man named Crowley bought a claim on Little Skookum Creek from a Mrs. Jeandreau. In July and August of 1898 there was controversy over the claim, and all parties agreed to have it surveyed. The surveyor did his work, but Nellie wanted to make certain, so she approached Mining Inspector Norwood. Now Nellie was flying by the seat of her trousers, and she used rumor, deduction, and wish-fulfillment to assure that Norwood would favor her in his decision.

Nellie had heard that Norwood had "a thing" with hotel owner Belinda Mulrooney in town, so she let it be known to Inspector Norwood that "someone in town" would be taken care of. And, just to make certain, she visited Belinda Mulrooney and talked vaguely of "her friend" the inspector, that things would work out fine, and so forth. What was happening was that Nellie Cashman was trying to grease the palm of Inspector Norwood — by giving money to his supposed good friend Belinda Mulrooney.

The whole affair led to a nasty inquiry, and Land Commissioner William Ogilvie personally took charge, went to Nellie's cabin, where she was recovering from illness, and asked her detailed questions. Ogilvie later wrote that "I was somewhat surprised. Miss Nellie Cashman, who, by the way, bears a very good reputation in this country." As soon as Nellie realized she would be required to substantiate her statements about Inspector Norwood "she began to beg off, pleading with me that she knew nothing about the matter, that she had been misled, etc." The pages of the government report are intriguing; below are a few portions, which give the flavor of the investigation and show how Nellie was forced to squirm under direct questioning:

Q. Why to Miss Mulrooney?
A. Because this is the proposition I made her, I did not make it to Capt. Norwood.
Q. Did she tell you she was acting for Capt. Norwood?
A. She did, but she couldn't.
Q. The facts are, that instead of Capt. Norwood making an offer to you, you made an offer to him?
A. No, not directly.
Q. You did when you asked him if it would change his opinion?
A. Well, we did not settle it.
Q. Well, you admitted?
A. Admitting and swearing to it is a different thing.
Q. Now, you state you made a suggestion?
A. No, just in behalf of her.
Q. I don't understand?
A. I asked him if it would change him if I gave her an interest in the claim.
Q. Why did you say her.
A. Well, I thought she had influence and would use it with him.
Q. What reason had you for thinking so?
A. Because I heard many people talking about it. You can hear anything, but you can't vouch for them all.
Q. You went on heresay?
A. Yes.
Q. So you would say, that even on her part, no effort was made to influence Capt. Norwood?

A. No.
Q. The only attempt was made by yourself when you asked him if an interest would influence him any?
A. Yes.
Q. You say there was some common report about Miss Mulrooney and the captain. Have you any reason to believe that Capt. Norwood had ever been receiving interests in claims through Miss Mulrooney?
A. Nothing except from rumours. If you are a saint in Dawson, you would be made a devil before you got through; if you carry a Bible in your arm you are a fraud. Common sense teaches that. I have been in mining camps for thirty-five years, and if I had common sense, I would not have bothered Miss Mulrooney. I think she was putting up a job on me.
Q. Do you know personally that Miss Mulrooney and Capt. Norwood were ever concerned in speculations of that kind?
A. No, I could not swear to it.
Q. Then, all you know, then, of the relations between Miss Mulrooney and Capt. Norwood, is based on rumours?
A. Yes, on rumours that I heard.[24]

Nellie was indeed off to a bumpy start in the Yukon. Inspector Norwood had an unblemished reputation, and her interrogator, Land Commissioner Ogilvie, was a man of high integrity. So not only was Nellie embarrassed and had to back-track, but she had also been bested by another woman, Belinda Mulrooney. Nellie Cashman may have been a known presence in the Yukon, but Belinda Mulrooney was one of the giants.

Nellie wasted no time, and on January 24, 1899, filed for a placer grant near Monte Cristo Gulch, adjacent to No. 29 Below, Bonanza Creek, practically on top of the previous, controversial claim.[25] This was a bench placer, meaning that instead of being in or near the river bed, it was on higher ground, having been created by ancient rivers.

Once more Nellie got in a legal hassle. William Thompson took his case to Gold Commissioner E. A. Senkler, claiming that

Nellie had actually encroached on his claim, known as the "Dickey hillside." In the meantime, Nellie kept working the claim, hired a laborer, and apparently began encroaching on a neighboring claim owned by M. E. Russell. She later proved that in addition to paying $1,500 for the claim, she had put at least $4,000 into the workings.[26]

Ill fortune plagued Nellie again, as Commissioner Senkler ruled in favor of Thompson and Russell.[27] Nellie being Nellie, she appealed to Commissioner Senkler, and hired the law firm of Belcourt & McDougal to represent her further. In a flurry of letters and statements over the next two years, Senkler grudgingly admitted that Nellie was doing what she thought was proper when she purchased and worked the claim, and then Senkler got involved in departmental puzzlement. His colleagues believed Nellie was due compensation; Senkler believed not. After all, an honest mistake had been made by all concerned, but why should the government pay for mistakes?[28]

A compromise was suggested, through the efforts of Nellie's attorneys and C.S.W. Barwell, surveyor for the City of Dawson. Barwell knew of vacant land, on Soap Creek, and believed that it was "fit for placer purposes." Barwell said Nellie's previous two losses — those of 1898 and 1899 — cost her ground worth $20,000 and $10,000. The Soap Creek claim was worth at least a few thousand dollars; let Nellie have it, stated Barwell, in a deposition of October 10, 1900.[29]

Commissioner Senkler called Nellie in to his office and took her deposition on July 8, 1901. A few of the statements follow:

> Q. Now what difficulty did you have after you purchased this property?
> A. I sent up a man to go to work on it, to sink and this man Russel, who claimed it too, interferred.
> Q. What did Russel claim?
> A. Russel claimed half of the claim. He claimed I was on his ground.
> Q. So that you still own half of the claim you purchased?

A. Yes.
Q. But the loss you suffered was through the result of his claiming half of the ground?
A. Yes, it took away all the pay-streak.
Q. Was that after you purchased?
A. Yes.
Q. What money did you take out of your ground?
A. Nothing, I never took a cent out of it.
Q. How much did Russel take out of his claim?
A. $20,000 out of the piece he took away from me.
Q. How do you know that?
A. I seen it.[30]

Probably wishing to be rid of Nellie, Commissioner Senkler ruled on November 30, 1901, that Nellie Cashman should receive as compensation ten claims on Soap Creek, which was a tributary of Gold Bottom. He specified that she must accept this as "full compensation," wishing the matter to come to an end.[31]

There is some evidence of a different court action in September of 1902, although case records are missing. Nellie sued a man named Jones in the Territorial Court for $700. This was a partnership dispute of No. 21 Below, Bonanza. Jones tried to get out of a contract, claiming that Nellie had signed in his stead. The court ruled otherwise, and Nellie was granted the $700, and had a good chance of getting an additional $1,000. A local newspaper ran a brief article entitled "NELLIE CASHMAN GETS THE MONEY," and concluded with a quote from her:

> I was confident of being able to show the truth of my statement that the supplies and provisions that were sent up there were all mine, and I showed it to be true.[32]

If Nellie Cashman made any money in the Yukon, it was at a claim she had in partnership with five others, at No. 19 Below, Bonanza. Clum and others report that she made as much as $100,000, and she may have done so. Fred Lockley, one of her more reliable interviewers, quotes her as follows:

This is a view of Bonanza Creek in 1897, its first big year. Nellie's best claim was No. 19 Below Bonanza, meaning 19 claims below that of the first discovery. (Natural History Museum, Los Angeles)

> I got hold of a claim — No. 19 below Bonanza. It proved to be a rich claim. I took out over $100,000 from the claim. What did I do with it? I spent every red cent of it buying other claims and prospecting the country. I went out with a dog team or on snowshoes all over that district looking for rich claims.[33]

She did move from Dawson down to No. 19, Below. In a 1902 directory she is listed as "Cashman, Nellie, grocer, 19 Below Bonanza." In the same directory are two Tom Cunninghams, one a cook, the other a miner. No address is listed for either. Because of their names and occupations, either one could have been Nellie's nephew, who stayed with her during most of her Yukon period.[34]

Until 1904 Nellie mined in the Yukon, and in addition to the claims at No. 19, Below, she prospected and invested in many other claims, some of them hundreds of miles from Dawson. She still maintained a restaurant, and divided her time between her two occupations.

Dawson by this time was almost a metropolis, one of the largest cities in the northern part of the world. This was a city of tens of thousand residents in and nearby, with four newspapers, three hospitals, three schools, forty restaurants, three theaters, nineteen laundries, thirty physicians, twelve dentists, five assay offices, four churches, and such a large number of saloons and brothels that the publicists avoided the figure; Dawson also had two bottling works, which in the large mining community did a thriving business.[35]

Although Dawson was a large place, and a lively camp with enthusiastic boomers, stampeders, prospectors, and speculators, Dawson was not a violent place. British law and Canadian policy meant a strict enforcement of gambling, alcohol, and firearms regulations. For its size and the type of inhabitants, Dawson had an impeccable record. The Northwest Mounted Police, few in numbers, had attained the reputation that kept frontier roughnecks from California, Montana, Arizona, and Nevada in check.

This did not mean that Dawson had no problems. The severe

weather caused many deaths. Furthermore, the city was built instantly, with little thought to traffic patterns or sanitation. Although Dawson had physicians, nurses, and hospitals, the city did very little to prevent people from having health problems. Flu, colds, and scurvy were commonplace.

An example of this disinterest in public health occurred in late spring of 1899. Dawson government officials ordered all garbage in town to be collected and placed on the ice in the river. With the spring breakup, tons of Dawson garbage went flowing down the Yukon River on ice, an example of a community solving its problem by giving it to someone else.[36]

Nellie Cashman, that hardy prospector, traveler, and wanderer to deserts and glaciers, was also affected by the tough environment. Complete records exist for St. Mary's Hospital. Nellie was admitted as a patient five times between 1900 and 1903. These were brief stays except for November 1902-February 1903, when she had some intestinal surgery. Even the bills and receipts exist for this institution. In September of 1902 Nellie spent two days in the hospital, at $5 a day.[37]

Unlike other Dawsonites, though, Nellie Cashman considered St. Mary's Hospital her home. From her first arrival in town she had helped in fund drives for the church and hospital. Nellie was a staunch Roman Catholic, and the fact that people of her religion were providing needed care for people in this mining community was cause for her constant support.

Although in Pioche, Tombstone, and Kingston, Nellie had done what she could to support the church, Dawson gave her an opportunity she had never had previously. Here was a community of priests and sisters with whom she had frequent contact. These religious people became Nellie's main social contact. Sometimes in the evenings, or on a weekend when she had spare hours from her restaurant or working her claims, she would go to the church and the hospital. There she would join the Sisters of St. Ann in serious talk, in humorous asides, in good fellowship. The following words of that era are typical:

> The Sisters of St. Mary's Hospital knew her well. With their knowledge of her true worth, her charity, her purity of life, her faith, her deep Catholicism, they ranked her among living saints. After spending an hour with ready pleasantry and Irish wit, she would say to the sisters: "Now I must go to the Chapel." And there she would remain an hour or more.[38]

There are, no doubt, other people about whom similar fine words have been written. Yet, moving in two dissimilar environments was a challenge few undertook. Nellie could work with the most crude, uneducated, unkempt mining crew, walk the streets of Dawson with the roughest toughs and prostitutes, serve meals and drinks to surly drunks and high rollers; yet she maintained her poise and self-respect. When the opportunity arose, Nellie sought the companionship of sisters and priests, united in their common goal of improving the lot of their fellow man. This was not typical behavior on the American mining frontier.

After getting hefty gold profits from her claim at No. 19, Below Bonanza, Nellie spent much of 1902 and 1903 out prospecting. She went into the Alsek country and bought claims on Bullion, Sheep, and Madeline Creeks. Near Bullion Creek she was involved in an incident which writers have expanded into hilarious frontier humor. Supposedly, she was in a roadhouse in a three-tiered bunk; on top was an obese miner, Nellie in the middle, and a "squaw" on the bottom. The bunk collapsed, and the miner and Nellie crashed on the squaw, with Nellie later stating, "I can still hear that squaw shrieking."[39]

The actual event was explained by Nellie to a journalist from the *Dawson Daily News* on January 26, 1904. Nellie reported that she had walked twenty-one miles alone on a trail in temperature of 61 below zero:

> Are there roadhouses on the Bullion Trail? Well, I should say. One night I occupied the top bunk in the fifth story of a series. The bunks were put up more for catching quick

Early view of St. Mary's Hospital and Church, Dawson. The hospital improved dramatically in a few years, thanks in some part to Nellie Cashman's fund drives. (Archives, Sisters of St. Ann, Victoria)

> transients trade than for stability, and they collapsed under the heavy superimposed weight of humanity. Ah, Lucky day. I came down on top. The bottom victim is there yet.

Towards the end of 1903 Nellie needed a rest, and decided to visit the various Cunninghams in Arizona. She arrived in late October and made trips to Flagstaff, Tucson, and Bisbee. Nephew Mike Cunningham was by now cashier in the Bank of Bisbee, thanks to the start Nellie had given him by placing him with the Copper Queen Mining Company there. A niece, Mamie, had married Miles Archibald who was a guard at the Yuma Territorial Prison. After a short stay, Nellie then left for a brief visit to St. Paul, Minnesota. There she visited with another niece, who was in a convent.[40]

ST. MARY'S HOSPITAL

(In Charge of Sisters of St. Anne.)

The Best Equipped, Best Heated and Best Ventilated Hospital North of Seattle.

Special Accommodations for Private Patients!

STAFF PHYSICIANS:

Drs. W. T. Barrett,
Alfred Thompson,
J. A. McArthur,
R. R. Macfarlane,
J. A. Sutherland,
J. O. La Chapelle.

ADDRESS:

SISTER SUPERIOR ZENON,

DAWSON, Y. T.

The next few months Nellie spent in a flurry of travel. She was back in Dawson by late January of 1904, went directly to the Alsek district to prospect and mine, and in the next two months made several trips from Whitehorse to Dawson, this time by train rather than dog sled or boat. The Alsek and Kluane Districts, where she was working, were to the west of Whitehorse.[41] Nellie was now sixty years old, and such activity in such a climate attracted nationwide attention. The New York *Times* of

February 28, 1904, printed details of her Yukon-Arizona-Whitehorse ventures and referred to her as a "Klondiker with as much energy and enterprise as any of the men who ever went over the trail." The following quote helps explain why she was there:

> Interviewed at Dawson, Miss Cashman declared that the dry, bracing frigidity of the Arctic Circle was more attractive to her than the sunshine of Dixie Land.

In spite of making such positive statements, Nellie did not believe Dawson was necessarily the place she wanted to be. After all, this was not just gold mining, it was gold placer mining. The good material was quickly gathered, and with tens of thousands of miners and prospectors in the vicinity, the end would come soon. The gold production for the Yukon for nine years was:

1896	$ 300,000
1897	2,500,000
1898	10,000,000
1899	16,000,000
1900	22,275,000
1901	17,368,000
1902	11,962,000
1903	10,625,000
1904	9,413,000[42]

This, of course, is not the kind of data that prospectors and miners have; the above chart was published years later by government officials. There was, however, a steep drop in production from 1900 through 1904, and local miners, merchants, and other residents knew what was happening, even if they didn't have the statistics. Less gold meant less food, less whiskey, fewer concerts, fewer newspapers, fewer employees, lower school enrollments, and so forth.

Dawson had peaked. Not only was production drastically falling, but the tremendous gold placer finds on the beaches of Nome had drained hundreds of Dawsonites. Wilson Mizner, Rex

Beach, Tex Rickard, the Murphy saloon brothers, and other leading lights had packed up for Nome. "Arizona Charlie" Meadows had even tried to float his saloon-theater from Dawson down towards Nome, but without success.

The year 1904 would be Nellie's last on the Klondike, but she never turned to the beaches of Nome, probably because Nome was too old — that boom had erupted in 1899. A newer gold rush appealed to Nellie, on the banks of the Chena River south of the Yukon River, where the new settlement of Fairbanks was rapidly becoming the leading town in the territory.

Nellie's time in Dawson had been not only profitable, but also very satisfying because of the success she had in working with St. Mary's Church and Hospital. The hospital boasted of being "the Best Equipped" north of Seattle.[43] In the Archives of the Sisters of St. Ann in Victoria is a comprehensive account of the founding and building of the Dawson church and hospital. Among the *bienfaituers de la Maison* [benefactors of the order] are priests, judges, physicians, members of parliament, and one female stampeder: N. Cashman.

One incident from Nellie's Dawson days was retold by her several times in interviews and says something about her values and sense of humor. She stated many times that she was an "American" even before she left Ireland. There was, of course, nothing anti-Irish in her sentiments, but like all Irishmen she felt a deep antagonism towards England, Ireland's master. Apparently, though, Nellie did not suffer any inconveniences in the two periods of her life when she was again under English influence, in the Cassiar District of British Columbia, and later in Yukon Territory.

According to Nellie, there was an English holiday being celebrated in Dawson and a burly English miner walked into her grocery store and asked, "I say, Nellie, where's your flag?" The rest of the conversation went this way:

> "Outside," she tartly replied.
> "Beg pardon, but it's not. I didn't see a Union Jack as I came in."

These men are panning in their cabin. Winter did not mean that mining ceased. Many miners used this time to work their gravels, or to rework some of their earlier diggings. (Archives, Sisters of St. Ann, Victoria)

> Nellie, unceremoniously, grabbed him by the arm, marched him to the porch. She pointed to the Stars and Stripes hung from the wall, and several English postage stamps on which were the Union Jack, holding it up.
> "There's my flag," she snapped, "and there's yours."[44]

An Italian prospector and miner, Felix Pedro, struck gold in the summer of 1902 on the Chena River, a tributary of the Tanana. Pedro's three years of looking paid off handsomely. The rush of the next two years led to two instant communities, Fairbanks and Chena, which was at the junction of the Tanana and Chena. In 1903 Fairbanks became the favored site, as newly arrived Judge James Wickersham placed offices of the Third Judicial District there. Before the end of the year about 2,000 people were in the vicinity, and Alaska's newest boom town was in full blast.[45]

The mission at Nulato on the Yukon River, operated by the Sisters of St. Ann, was one of Nellie's favorite stops. (Archives, Sisters of St. Ann, Victoria)

By early 1907 Fairbanks, though primarily still a wooden town, had a population of 3,500, with about 12,500 in the surrounding district. A narrow-gauge railroad, two hospitals, five churches, twelve hotels, eleven physicians, fifteen saloons — these were some of the hallmarks of the town's growth.[46]

When Nellie left Dawson for Fairbanks in mid-1904, she immediately opened a grocery business and miners' supply house. She had arrived in town at precisely the right time, and by the end of the winter of 1904-05 had a profit of $6,000.[47]

The fledgling mining community of Fairbanks needed a medical facility, and early in 1904 a one-storey hospital was built, under auspices of the Episcopal missions. Deaconess Clara Carter and Miss Annie Cragg Farthing were in charge, and they were strapped for funds, equipment, and just about anything a hospital needed. To the rescue came that experienced mining camp strong-armer,

Nellie Cashman. Nellie was a dedicated, devout Roman Catholic, but was ecumenical and broadminded. She admired the work of the two women trying to operate St. Matthew's Hospital, so she did what she had done countless times on the frontier. She toured the gold camps along the Tanana and other rivers, going by dog sled, boat, or on foot, and at any gathering of miners, whether in saloon or in the creeks, Nellie made her pitch for the hospital fund.

> Okay, boys! This is for the hospital. You've all had the good of it, you low down blankety-blank varmints — and if you got money to throw away at poker, you can give it to them hard-workin' Christian women that's takin' care of the sick.

These are words of a medical missionary who knew Nellie in this period, who described her "weather-beaten, freckled face creasing into a maze of wrinkles." Nellie obtained hundreds of dollars for the Fairbanks hospital, and "a more colorful and entertaining highwayman never mushed the trails of the Yukon."[48]

In Nellie's correspondence, documents, and interviews, Fairbanks does not loom very large, even though some accounts place her there from 1904 through 1907. The only noteworthy activities were the great financial success of her first year there as a merchant, and the pleasure she had from heading the fund drives for St. Matthews Hospital. The reason for her lack of interest in Fairbanks was because her thoughts and plans were already elsewhere. She had heard rumors and received some first-hand reports of placer gold far to the north, hundreds of miles above Fairbanks, above the Arctic Circle.

Access to that country was via the Koyukuk River, which although in the far north, was reached by going hundreds of miles to the west, first down the Tanana River to the Yukon River, then to the Koyukuk. In July of 1905 Nellie made her move, the move of her life. She was heading into the Koyukuk country, which was to be her true home, where she would remain for twenty years.

Her last fling with civilization was a stop at Nulato on the Yukon River, where the Sisters of St. Ann operated St. Peter Claver Mission. The Mission diary entry for July 8 compresses many facts and attitudes in a few words:

> Arrival of Miss Nelly Cashman. She took dinner and supper with us; gave $10 in cash next day; sent half-crate of oranges, onions, potatoes, steaks, roast beef, candy, etc. She made her Easter duty also; and what was the most amusing was to see her a few days after her arrival make all the Jesuits fly to catch a big log which was floating by.[49]

On July 14, 1905, Nellie headed up the Koyukuk. In late May of 1906, the town of Fairbanks was engulfed in flames, and almost every building in the business district was lost. This disaster had little impact on Nellie Cashman. She had spent her winter in Fairbanks and earned a fancy profit, enough to enable her to go on to further prospecting and mining, which to her was more interesting and challenging than selling groceries or mining tools.

The Koyukuk was a main tributary to the Yukon River. Nellie's mines at Nolan Creek were slightly above where the Koyukuk intersects with the Arctic Circle. (Century Magazine, 1890 – 91)

VII

FULFILLMENT ON THE KOYUKUK

ALTHOUGH most accounts indicate that serious mining on the Koyukuk River Basin began around 1910, in reality considerable prospecting and mining had taken place for a few decades. John Bremner, Gordon Bettles, and N. V. Hendrick had all tried their luck before the mid-1890s, and following the Klondike stampede of 1896-98 the entire northland was visited by prospectors, some of whom went up the Koyukuk and its three forks.[1]

Geologist Josiah Spurr estimated that in 1896 the population of the Koyukuk was twenty-five, and that at least $5,000 in gold had been removed. Geologist F. C. Schrader visited the country in 1899, and he believed that at least 1,500 people were in the basin prospecting and mining. In other words, from the beginning of the Klondike Rush, some adventurous prospectors decided to consider other likely spots in the north country.[2]

One interesting account is by Britisher J. Hudson, who with a few friends went up the Yukon River in 1898-99, then decided to explore and prospect up the Koyukuk. The expense of provisions, severe cold, desolation, and about fifteen other factors made the party turn back. What remained in Hudson's mind was a frozen foot, a chap who was accidently shot, "several men lost their wits," and others who got scurvy. He entitled his Koyukuk account "Our Klondike Failure."[3]

The Dawson and Nome newspapers belittled the Koyukuk

efforts, trying to discredit the gold placers so as to prevent "outflow" of their populations. One tale reported "terrible suffering and privation" by a Koyukuk prospecting party; they didn't find any gold, were bothered by the Indians, and "were all used up." Another report suggested that the Koyukuk riches were concocted by Gordon Bettles, who had settled in up the Koyukuk and operated a little store. He wanted customers, who would arrive if they believed gold awaited them.[4]

The Koyukuk River is over 550 miles long and is one of the main tributaries of the Yukon River. The Koyukuk Basin was bordered on the west by the Seward Peninsula, in the north by the Brooks Range, and on the southeast by a massive uplift that trends diagonally across central Alaska. There was no easy way to get to the Koyukuk River. From the Pacific Ocean, one had to travel 450 miles inland up the Yukon River, before reaching the Koyukuk. To really get into the real Koyukuk country, one had to take a small steamboat another 450 miles north to the Episcopal mission of St. John's-in-the-Wilderness at Allakaket; then, depending on how high the water was, a small craft could go an additional 80 miles north to the trading post of Gordon Bettles.

If that itinerary sounds dreadful, time-consuming, and tortuous, consider the fact that upon reaching Bettles one was nowhere near the gold country yet. The best, most consistent gold placers were at Wiseman and Nolan Creeks, on the Middle Fork of the Koyukuk, eighty miles north of Bettles. This was just north of the Arctic Circle and was to be Nellie Cashman's home for twenty years.[5]

The Koyukuk River was large but not particularly deep, and the current averaged about four miles an hour, sometimes less. There were few rapids, so the river was ideal for flat bottomed steamboats. The entire basin was about 150 miles wide, and the valley in general was broad and open, with some gently rolling plateaus. The Koyukuk system drained the entire northwestern part of the Yukon Basin.

This was big country, but contained very little, only sparse vegetation, small, scattered groups of spruce and other evergreens, very few mammals, and even fewer people. By 1900, in the entire basin, there were only a few hundred native peoples, both Eskimos and Indians. Add to that the several hundred prospectors and miners, and the total for the basin of thirty thousand square miles was a human population of less than 1,000.

Episcopal Deacon Hudson Stuck is one of the more reliable chroniclers of this part of Alaska early in this century, and he was fascinated and perturbed by the Koyukuk country. He used such phrases at "utmost monotony" for much of the landscape, and stated that the chief impression of the region was its loneliness:

> More than once I have journeyed three hundred miles up this river without seeing a living soul, white or native.

Sometimes, he believed, it was better not to meet people. In spite of low population, there was scattered violence, as the Indians on the Lower Koyukuk had a reputation for truculence. Several shootings of prospectors had taken place before 1900, and in 1912 the "Blueberry Kid" murdered "Fiddler John" and "Dutch Marie," who were returning from Nolan Creek with their gold stakes. In general, though, even the Indians did not like the Koyukuk country, as there were few animals, fish were not plentiful, and the cost of "white man's grub" was high.[6]

Although this area is remote and relatively uninhabited to this day, geologists have done considerable work in the region, often combining the Koyukuk with the huge Chandalar [or, Chandlar] Basin to the east. The region contains both placer and lode gold, but the overwhelming bulk of the gold production — 95% — has been from the placer fields. The gold-bearing belt is roughly eighty miles long and runs in an east-west direction from the Chandalar Basin through the Koyukuk Basin.[7]

There seems to be no consensus as to the origin of the placer gold. One important study maintains that the placer gold is obviously from the local lode deposits that have broken down; there are considerable spatial, mineralogical and geochemical similarities between the lode and placer deposits.[8] Another recent study, though, maintains that "lode sources have not been found for the majority of placer deposits in the Koyukuk or Chandalar districts."[9] Well, prospectors and miners then and now were not particularly concerned with such theorizing. Placer gold was found, especially by and in the streams along the Middle Fork of the Koyukuk River.

This was the last chapter in Nellie's life, and it was a long, difficult one, but packed with adventures and challenges. For her, and for the Nolan Creek area generally, the time is logically divided into three periods. The first was 1905 through 1912, when the initial serious work took place in the area. The next phase lasted until around 1919, when important placers were located and worked and sizeable gold production was realized. From then until her death in 1925 Nellie was readjusting, trying to find additional financing in order to apply larger machinery and better techniques to her claims.

As we recall, Nellie arrived in Fairbanks and did very well in her grocery supply business the following winter. The fellow who found gold in 1903 on Nolan Creek was O. R. Williams, and he named the place for John Nolan, a shadowy figure in the region. Williams then filed Discovery on Nolan Creek. On July 27, 1905, Nellie showed her first legal interest in the Koyukuk country, as she obtained a lease on "Fay Pup" on Nolan Creek from Volney Richmond, who was general manager of the Northern Commercial Company in Fairbanks, Nellie's main supplier. Richmond was keen on the area and apparently passed on this enthusiasm to Nellie. She then left to spend her first winter on the Upper Koyukuk.[10]

In the next few years Nellie was involved in many mining transactions and some claims she worked; others she merely sold to someone else. Most were around Nolan Creek and included the Cashman Bench, Eureka Claim, Bunker Hill Bench, Sweet Marie Bench, Sullivan Bench, Fay Gulch, and others. She filed a

Nolan Creek in 1913, where Nellie worked claims for twenty years. (U.S. Geological Survey, Bulletin 532)

few claims on Big Creek in the Chandalar District to the northeast in 1906-07 but was never very interested in that district. The best clear statement of Nellie's properties was by J. W. McCord, a longtime prospector-miner in the area:

> The properties of Nellie Cashman, consisting of eleven placer claims, situated on the right limit of Nolan Creek, are all immediately adjacent to and surrounding the richest placer deposits that have been discovered in the Koyukuk district.[11]

Although Nolan Creek was far from everywhere, news about the district did get around, and even the press in the distant Yukon kept its readers informed on Nellie Cashman, "the pioneer Dawson woman trader and miner." The *Dawson Daily*

News of October 1, 1906, stated that Nellie had been rediscovered. The trading post manager at Bettles reported that Fairbanks had been "too tame for her and she pushed on to Koyukuk." Nellie was sinking a shaft on Fay Pup, looking for bedrock. She believed that the bedrock would be "false bedrock" and more depth and penetration would be needed to find the true source, the deposit of placer gold.[12]

One must keep in mind the remoteness of this region when considering gold mining. At this time the area was not only remote but was the farthest north gold placer operation in the world, more so than even the Siberian placers. Because of the enormous costs of transportation, ground that would be considered very rich in other districts might only be marginal when the exorbitant expenses were determined. Nolan Creek would prove to be a very rich placer field, but profits were modest because of the costs.[13]

An encounter in the winter of 1907-08 provides us with a first-hand account of Nellie. Nineteen-year-old Clara Heintz, from Los Angeles, was an assistant at the Episcopal mission at Allakaket, St-Johns-in-the-Wilderness, some four hundred miles up the Koyukuk River. One winter evening a sled pulled up, and a figure in man's clothing — pants, fur parka, boots, and heavy gloves — knocked on the door. Young Clara was mystified at what followed:

> As Miss Carter opened the door, I heard her shout, "Nellie!" while Nellie cried, "Deaconess!" and the two women embraced, with the little Deaconess almost disappearing in the frosty bear hug of the visitor. Miss Carter introduced me to the stocky middle-aged woman with graying red hair whose weather-beaten face was a battlefield of freckles and wrinkles.

This strange visitor looked over the room, took in every object, and would comment: "Well, I'll be goldurned! If this ain't the most. . . ." Tea and biscuits followed, with young Clara hanging on Nellie's tales of the desert and glaciers. Clara's reason for Nellie's spinster status: "Seen too much o' the worst side of men, I guess." What the young California girl remembered about Nellie was her kindness, her toughness, her rough, musher's jargon, "but there was never a breath of scandal connected with her."[14]

Gold being what it is, the word of success at Nolan Creek gradually hit all the mining camps in Alaska. People were interested in the strike, but distance and climate kept most of them away. In the winter and spring of 1908, the Chena *Tanana Miner* carried more than a dozen stories on the Upper Koyukuk mining, especially at and around Nolan Creek. The issue of February 23 was particularly important, because it contained a brief statement from Frank E. Howard, U.S. Commissioner of the Koyukuk District. His comments left no doubt as to the future of the field:

> I think this is now the best and most open field in Alaska for the prospector; at least, considering what little prospecting has been done up to date, good results have been obtained. In one hole on Nolan creek 80 feet of a very rich paystreak has been uncovered. Other holes are now going down along the same line, and the whole camp now looks forward to the best results.

In the summer of 1908 Nellie went to Fairbanks on business and was interviewed by several of the newspapers. She explained that the prospects on Nolan Creek were good. The town of Coldfoot had been pretty much abandoned, and a new settlement was begun at the intersection of Wiseman Creek and the Middle Fork of the Koyukuk. This would be the center of trade and housing for the upper Koyukuk for decades to come, and was known variously as Nolan or Wiseman.

Nellie was in Fairbanks to purchase boilers for herself and for other miners. Good pay had just been struck on Nolan Creek, and Nellie stated that "there is no doubt but that that camp is going to be a good one." Boilers of varying sizes were used underground. The steam heat from the boilers would be concentrated in long tubes with thaw points at the end. The points, touching or inserted into the permafrost, would thaw the ground, enabling the miner to remove the ground and take it to the surface, for working then or later.[15]

However, this was no typical miner in town but "the well known sourdough woman prospector," so many questions were asked of her related to her personal life, her philosophy, and the future. She enjoyed this kind of banter and made good copy. She intended to strike it rich, and if so "I will put it back into the ground again — give some of the boys work in opening up new districts." She predicted greatness for Alaska, pointing out the positive changes already in places like Fairbanks, with its homes and permanent families.

She brushed aside questions about her spinster status, claiming she had "been too busy to talk about such things." However, when pressed about living with men in the wilderness she outlined her working code:

> I have mushed with men, slept out in the open, swashed it with them and been with them constantly, and I have never once been offered an insult. You won't find that class of men among the sourdoughs of Alaska. A woman is as safe among them as any and all times as at her own fireside. If a woman complains about her treatment from any of the boys, you can take my word for it she has only herself to blame. I have prospected in Wyoming and in other states of the West and am now getting to be an old woman, and I can truthfully say that there was never a bigger-hearted or more broad-minded class of men than the genuine sourdoughs of Alaska. God bless them.[16]

Crowded weeks of getting the machinery back to Nolan Creek were followed by a respite for Nellie. She went down the Koyukuk River, down the Yukon for additional hundreds of miles, and upon reaching the Pacific Coast, took a steamer for California. She went to Bisbee, Arizona, for a brief visit with her nephew Mike Cunningham, by now a well-established banker in Cochise County.

Miners cabins at Wiseman. (Marshall, Arctic Village)

Her return route was indeed far from routine. She left Fairbanks on the stage for Hot Springs, then took a stage from that point to Fort Gibbon. From there she hired an Indian laborer and a dog team and went overland to Coldfoot, then up the remaining twenty miles to her camp on Nolan Creek. While in Fairbanks Nellie had gotten a report on her claims, which were looking very promising. The dog team in the snow, cold, and darkness of late November going across the Koyukuk Basin earned the admiration of the *Fairbanks Daily Times,* which claimed that she still showed plenty of vitality even at her advanced age, "and the many years of life spent on the frontier has equipped her for the exposure she is certain to encounter in traveling in Alaska."[17]

The news Nellie had received of the golden promise of Nolan Creek was based on sound information. The winter of 1908-09 saw considerable work done on the many Nolan Creek claims, and the new boilers performed well. One small operation, by John and Gus Oleson, hit bedrock in the winter at 120 feet and

A native camp along the Koyukuk around 1910. Many narratives of the period do not distinguish between Eskimos or Indians. (Stuck, Voyages)

struck coarse gold pay ground; in three months they took out a profit of $105,000.

The same report also provided news of goings-on at No. 1, Below, owned by Nellie Cashman. Nellie and her small crew had also hit bedrock and pay gold, but she ran into problems. At the point she reached bedrock, the area was one of large granite boulders, which interfered with the lowering and raising of buckets. So, she had to sink an additional shaft to the bedrock in order to work the ground.[18]

During this winter there were more than 200 people in camp working claims, and additional cabins were being erected. This was difficult, as there was little timber nearby. In general, the upper Koyukuk had few trees, and the voracious appetite of the boilers meant that the miners had to go farther and farther for firewood.

A report from the district in August of 1909 mentioned that some of the gold of Nolan Creek was found along the benches, at a depth of fifteen feet; yet other gold was found much deeper, which puzzled the prospectors and miners. With such circumstances, one never knew what to expect, which could mean instant reward or time-consuming, worthless labor. This report also mentioned a Nellie Cashman find of note; she located a hot spring on No. 1 Below, "possessing medicinal properties." Nothing was later mentioned about this curious find, so at best Nellie obtained some water to use in her placer mining.[19]

The placenames Nolan, Nolan Creek, and Wiseman present confusion, as sometimes the names referred to particular locations; yet other times the names were used interchangeably. Nolan Creek was a tributary of Wiseman Creek, which went into the Middle Fork of the Koyukuk. In 1910-11, because so much activity was centered at Nolan Creek, the town of Coldfoot 15 miles to the south was abandoned. B. E. Wright kept a roadhouse at the junction of Wiseman Creek and the Koyukuk, and the place became known as "Wrights," later as Nolan, and also as Wiseman. In 1909 the postoffice was called "Nolan," but was changed to "Wiseman" in 1923. During these years, even though most people lived at Wiseman, there was considerable mining at Nolan Creek, where dozens of people lived in cabins.[20]

In this early period, 1907-1911, Nellie usually hired two or three men to work her claims, and much of the time they worked throughout the winter. This is not as silly as it seems. This was underground work, often at or seeking bedrock, which was many feet below the surface. Here the small crew could work in winter as well as summer, as permafrost was permafrost. The boilers created steam to thaw the permafrost, while providing heat for the underground crew. Thawed ground was raised to the surface, where it could be worked in the late spring or summer, when water was available from Nolan Creek.

Nellie owned most of her claims outright, but a few were co-owned. She was involved with R. E. McIntyre in several claims, and they would remain professional and personal friends until her death. Bob McIntyre was widely known in the district, well liked, in spite of his insistence at singing — off key — at all local functions. Another interesting name that shows up on claims and legal papers for Nolan Creek is M. J. Cunningham of Bisbee, Arizona. This Cochise County banker was probably not particularly interested in Koyukuk mining claims, but may have been persuaded by Aunt Nellie that "big pay" was just ahead, or he may have humored her with some casual investments. Mike Cunningham did remain interested in mining, but in Arizona. In December of 1909, while in Bisbee on one of her trips "outside," Nellie signed over to Mike the Last Chance mining claim in the Turquoise District of Cochise County. This appears to have been Nellie's last participation in any Arizona mining activity.[21]

There are no ledger books, no profit and loss statements for Nellie Cashman on Nolan Creek, although there is a fairly decent paper trail. She worked hard, always had a man or two on the payroll, made many outside purchases of supplies and equipment, and apparently never owed a dollar to anyone. J. W. McCord, a leading prospector-miner in the district, put in writing a statement regarding Nellie's finances and plans, which included this endorsement:

> Miss Cashman herself is known by every old-timer in the North, both in Alaska and the Yukon Territory. She has never made a failure, either in her intentions or in the payment of her bills. These facts are vouched for by the biggest trading companies in the North, and by Louis Sloss himself, head of the Northern Commercial Company and the Northern Navigation Company, the largest companies operating in Alaska.[22]

The U.S. Geological Survey kept accurate production and activity records for the entire country, and the few comments on the Koyukuk District for 1914 indicate a few of its peculiarities and problems. Gold produced was $260,000, down from $400,000 the previous year. The decrease was due partly to lack of water, but mostly due to a few abnormally rich placers in the Hammond Creek region in 1913, which were quickly exhausted. For a number of years the Koyukuk had averaged between $200,000 and $300,000. In 1914 there were about thirty mines in the district, and most of these were in the Wiseman Creek-Nolan Creek areas. Half of the mines were worked in the summer, and half were worked in the winter, but four-fifths of all production was from those operating in winter.[23]

The upper Koyukuk would never have a true gold rush, and by 1914, although few realized it, the good years were in the past. Nolan Creek, Hammond Creek, and other areas would continue for thirty years to provide some gold, but it was with hard work, at great expense, and never with rich finds or great hopes for the future.

Mining engineer Emil Hurja examined many properties there in the summer of 1914 and used phrases by now commonplace: "remoteness," and "difficulty of access." He reported on each area and stated that the claims around Wiseman Creek were the deepest placer shafts in Alaska: one was 280 feet, another 335 feet. Nolan Creek had done well in the past "but very little work is being done there now." The total population for the entire upper Koyukuk was about 200.[24]

By this time an important work had been published by the U.S. Geological Survey by A. G. Maddren. This was a clear examination of geology, mining techniques, and history, and contained some detail on most major workings. Maddern even included photos that dealt with transportation and climate in this area north of the Arctic Circle. He described Nolan Creek as about four miles long, heavily mixed with glacial silt, with the gold placers of many types: bench gravels, shallow gulch gravels, and deeply buried frozen gravels. The best gold seemed to be in the deep frozen gravels, which he maintained were first worked successfully by use of boilers brought into the country in 1906; these were arranged for by Nellie Cashman. The best of these deep placers were from Discovery to No. 6 Below, a distance of less than two miles. Nellie owned and worked two of these claims from time to time.[25]

There was a good production spurt for the next few years, and one mining journal mentioned in 1915 that Nolan Creek had produced a number of small fortunes in the past eight years but was considered worked out; ". . . it is showing up well this season."[26] The previous winter had an unfortunate weather quirk that is difficult for outsiders to comprehend. The weather was "not cold enough" to suit the operators. The streams did not freeze early enough, which hampered transportation, which on Wiseman and Nolan Creeks was by dog sled.

In this brief spurt, which peaked in 1916, six or seven of the creeks did fairly well; some were old producers, and a few were newer operators on adjacent or nearby creeks. The gold production chart for the area showed that there was still some profit in the Nolan Creek beds:

Nolan Creek	$145,000
Hammond River	105,000
Myrtle Creek	11,000
Gold Creek	10,000
Jay Creek	22,000[27]

One may read the many geological reports and statements of gold production without finding many references of specific properties or owners. On the other hand, some deductions are possible. The above statements suggest that some fortunes had been made on Nolan Creek, and just recently $145,000 had been taken out by the operators.

Nellie Cashman was one of these operators, and she must have been doing fairly well. She decided to treat herself to another trip to visit the Cunninghams in Bisbee, Arizona.

By scow and steamboat she made the long trip down to the Yukon River, then went all the way up, to Yukon Territory. She visited old friends and familiar places in Dawson, then went down to Whitehorse. Here she boarded a train to Skagway, then took a steamship to Seattle and on to Arizona.[28]

In Seattle she was interviewed at the Hotel Northern, where she was described as a slight little woman, with gray hair brushed straight back from her forehead, and "big hands roughened and muscular from doing the work of men along the trail and in the camps." She glanced down at her skirts, which she described as a necessary inconvenience of her stateside visit: "These things will go pretty quick when I get back up there. Fine time I'd have with skirts on the trail."[29]

The many years that Nellie worked her claims on Nolan Creek were not just spent in sinking shafts, hauling ore, or looking for firewood. She lived in a small, active community, one of the most unusual in North America.

The people there were weird. This is how they defined themselves, and this is also the interpretation of others who knew them. An early appraisal comes from Episcopal Deacon Stuck:

> I think some of the best men that Alaska contains are to be found in the Koyukuk, and I will not say that some of the worst are not there also. It is commonly said, and I think with ground, that more whisky is consumed here per capita than in any other community in Alaska, and perhaps in the world.[30]

To Stuck, these "Old Koyukukers" were in Alaska's most interesting mining camp. Its isolation gave it a flavor unknown elsewhere.

Another early report on the residents came in 1908 from Episcopal Bishop Rowe, who went out of his way "to mush to the new Nolan creek strike." He stated that the people there were mostly old-timers of the conservative kind, "the type which has pioneered the way into this country for their fellow men, and who have the true spirit of the North. I do not believe you will find a finer lot of men in any community than those Koyukuk miners."[31]

A photo of Koyukuk oldtimers around 1930. All of these men knew or worked with Nellie Cashman. (Marshall, Arctic Village)

Mining correspondent Emil Hurja from San Francisco moved warily among these people. He reported that they "look with disfavor upon any influx of outsiders."[32] Of course, Hurja was indeed an "outsider," while Stuck had sledded and snowshoed in this region for years and was considered one of Alaska's oldtimers.

The strangeness, the unique flavor of life in the Upper Koyukuk was vividly portrayed by traveler Robert Marshall, whose *Arctic Village* was published a few years after Nellie's death. Marshall immersed himself in the Wiseman-Nolan creeks' way of life and enquired into every aspect of thought and deed.[33]

Whiskey "had the priority over everything else." The few hundred people were from all parts of the earth; Croatia, New York City, Arizona, Ireland, Sweden, and so forth. Some were born losers, and the Koyukuk country was a reasonable stopping place for failures or those nearing the end of the line. Others were attracted to the Koyukuk because they sought, in fact, thrived in its isolation, where a new way of life would be created by the few people who lived there. The people who came to the Koyukuk tended to stay, even when the gold petered out, or when other prospects were dim. They found a comaraderie there which provided either what they lacked, or what they grew to appreciate. One of the things that held them together was their divers activities. Every person on the Koyukuk could do practically everything. One had to be a blacksmith, a mechanic, steam engineer, logger, geologist, carpenter, baker, cook, dog musher, and laundryman. A shirker was out of it, not respected, not tolerated. Because of the many talents needed and rapidly developed, the Koyukukers had much to talk about, and talk was the main activity of the camps.

Marshall even gave percentages of the topics of conversations. He came up with a figure of 3,539 minutes of conversation for each individual he interviewed, and categorized the topics in the following blocks of minutes:

recreation (cards, drinking, etc.)	255
sickness, accidents, death	184
sex, from a personal standpoint	152
personal reminiscences	152
quarrels in the Koyukuk	105

For so few people in the vicinity, one may wonder about the discussions of "sex." The earlier period, ending about 1920, saw considerable prostitution in the Wiseman and Nolan Creek camps. A few Eskimo women, some Indian women, and the odd "Big Toothed Kate" or "Heavy Helen" discarded from the brothels of Nome or Dawson found their way to the Koyukuk. Sex was practiced, acknowledged, and was a common topic of conversation.

One of Marshall's most intriguing statistics was to combine Upper Koyukuk figures for population, gold production, and number of prostitutes. A few select years follow:

YEAR	POPULATION	GOLD	PROSTITUTES
1901	320	$173,000	6
1902	350	200,000	10
1907	120	100,000	0
1909	230	420,000	5
1911	160	130,000	5
1915	300	290,000	14
1920	119	90,000	0

Although Marshall included figures for each year, the above summary indicates the close relationship between production and the number of prostitutes, which should come as a shock to nobody.

There were many other unusual aspects of this Arctic camp. Crime was practically unknown, even though quarrels were frequent. There was a U.S. Commissioner, who had few duties. For clergy there were Mr. & Mrs. Pingel in camp, ex-missionaries, but they did not do much with the "God business" here.

The environment set this area of the world apart from where man tended to live. Consider the following causes of death in the Wiseman-Nolan creeks area from 1898 through 1931:

freezing	12
drowning	10
mining accidents	6
disappearing	6
suicide	5
freighting accidents	3
murder	3
burning, scalding	2
accident, intoxication	1
accident, self-shooting	1
overeating [after starvation]	1

This is a table of fifty deaths, most of them not common to residents of New York City, Linn County, Iowa, or San Diego, California.

The people in the Upper Koyukuk shared dangers and death

Nellie's occupations differed greatly from that of these prostitutes, yet she was in frequent contact with them and could count on them for charitable contributions. (Archives, Sisters of St. Ann, Victoria)

along with possible profits. In this isolated world they came to feel unique, important, that they could function and survive. But they could do so only because of the tight-knit community. In a seemingly contradictory way, they were all "my own boss," yet each individual knew he fit, or had to fit, into the community.[34]

This unusual environment, that so easily encouraged the use of alcohol, is to this day also associated with certain types of mental illness. The winter gloom, with its long, almost endless nights, took its toll. One of Nellie's acquaintances in this part of Alaska was Dr. "Hap" Burke, who had married young Clara Heintz from the Allakaket Mission. One of Dr. Burke's greatest worries was what to do with people in the winter. Not only isolated miners and trappers succumbed to the winter gloom, but some of his nurses were also treated for "hysteria, melancholia, or incipient paranoia."[35]

This is the type of environment where Nellie Cashman decided to spend the last twenty years of her life. To some, Nolan Creek was a camp of misfits, thrown out of the world into darkness and cold. Yet, they survived, some even thrived. They had frequent parties in the roadhouse at Wiseman, where a badly played violin accompanied Bob McIntyre's screeching voice, the crowd downing many whiskies. Nellie Cashman was there, for twenty years.

This was a tough crowd. Nellie was tough. These were men she worked with, some she hired, with some — like McIntyre — she had business arrangements. Hard work, quarreling, drinking, sex for sale — such was life on Nolan Creek. This was a frontier mining camp, but it was the toughest that Nellie had encountered. She remained what she had been, a hard-working no-nonsense businesswoman, taking people at their word. The fact that in such a crude environment Nellie and these hard cases co-existed for two decades forces the conclusion that they had built up considerable mutual respect.

VIII

END OF THE PAYSTREAK

GOLD production suffered during World War I, as many of Alaska's young men were called into service. This created direct and indirect impacts on the territory. The mines and mills had to get along with fewer workers; in fact, the entire economy had to do with less. Many of the hardy employees in the lumber camps, on the railroads, and in the river and overland transportation systems had been young men, and their loss was noticeable.

News of the Great War took months to find its way to the Upper Koyukuk, but finally in the winter of 1918 even Nolan Creek found out that the United States had entered the war. Nellie the patriotic at once spread the word throughout the region. She convinced several dozen young men to go the battle fields of France. Not only that, but she would lead them there. They left the Koyukuk and traveled several hundred miles through snow and ice to Fort Gibbon. However, at that post they learned of the Armistice, so the Cashman Battalion never left Alaska.[1]

On the other hand, even without the Great War, there would have been no bonanza days on Nolan Creek. The good days were in the past, but even Nellie Cashman resisted history. She and a few dozen operators and claim owners kept working, kept hoping, but Nolan Creek would only stumble along just as a marginal placer producer.

This last part of Nellie's life was still spent in dreaming and planning. Now she made almost annual trips "outside," sometimes to Fairbanks or Juneau, and on a few occasions went south to visit in Arizona. As far back as the period 1900-1904 she had been good newspaper copy, so that by the early 1920s she was among the legends. She began to get quite taken with herself, and in interviews over the years she would toss out witty sayings or bits of philosophy, such as:

> Never borrow trouble or draw interest on past mistakes. There is too much to do in scattering about sunshine. It is just as easy to look upon the cheerful side of things as to be looking on the shadows all the time.
> More people kill themselves from needless and senseless worries than because of any other reason.
> Don't throw more weight upon a burdened brother or sister, but give them a little lift with their load.[2]

In the summer of 1919, she was just back in Fairbanks from a trip to California and Arizona, where she had dental work done and couldn't get away fast enough from short skirts, high heels, and people who complained too much about the high price of corsets. "That baby stuff is not for her." She advised those Californians to "get out of there and come to Alaska where life was life and not one nose wiping exhibition after another."[3]

The winters, though, seemed to be getting to Nellie, as she often spent several months away from the Koyukuk country. For example, in December of 1920 she arrived in Juneau on business, intending to stay for a few weeks with her long-time acquaintances, the Sisters of St. Ann, who operated a hospital in Juneau. The local paper believed that she would go again to Nolan Creek to "take out another fortune from the paystreak there."[4]

Nellie remained in Juneau for a few weeks, and in one interview stated firmly that her career was far from over:

> Tell the boys that I can still take the trail and no one has ever passed me yet on a long mush. I still can drive the dogs

> and use the snowshoes and that no one will ever have the satisfaction of ever having to break trail for me. When I feel that I can no longer keep up I will keep off the trail, but that is many years hence.[5]

On January 14, 1921, Nellie boarded the steamer "Jefferson" and headed for Seattle. She intended to purchase mining machinery, even some hydraulic pieces, in order to try something new with her Nolan Creek claims.[6] Nellie took precautions, and to convince any hesitant bank manager or mining equipment officer she had with her a sheaf of personal testimonials. Cecil Clegg, President of the Fairbanks Republican Club, referred to her "great energy and persistence," and called her an "inveterate booster for Alaska." L. T. Erwin, U.S. Marshal for the Fourth Division of Alaska, described her as "the oldest and most distinguished Pioneer woman" the territory had produced. In addition to her mining success "she has ever administered to the sick and needy, always wearing the same smile of hope and cheer through sunshine and through shadow, in every camp where she has lived." She also received a statement from Daniel Webster, U.S. Commissioner and Recorder for the Koyukuk District, stating that there was "no incumbrance of any kind or nature whatsoever against the Nellie Cashman property holdings."

These were not second-rate commendations, but Nellie wanted to be thorough, so she also had a statement from James Wickersham. By this time Judge Wickersham was the most well known and respected politician in Alaska, and he gladly gave her letters of introduction to Seattle businessmen. Wickersham stated that Nellie "has been engaged at the mining game in Alaska for all these years, and knows it better than anybody." Wickersham wrote one letter on her behalf to W. D. Stephens, Governor of California, asking him to assist her if possible. "She is one of the most capable miners in Alaska, and a woman worthy of the highest credit."[7]

There is some obvious exaggeration in the above testimonials, but the statements were based on Nellie's many years of

This photo was captioned "Alaska's two Nellies," showing Nellie Cashman and Nellie Lawing, another well-known territorial personality. (Archives, Sisters of St. Ann, Victoria)

experience. Furthermore, not too many individuals could garner such support for outside financing.

In late 1922 Nellie Cashman planned to incorporate, with the idea of raising additional money in order to work the Nolan Creek ground by hydraulic methods. She organized, but did not incorporate, the Midnight Sun Mining Company. The stock certificate is a fascinating revelation about Nellie and about doing business on the Upper Koyukuk. The firm had "no offices" and "no officers"; the person in charge was Nellie Cashman, Trustee. The company was organized with 50,000 shares, to sell at $2 each. What may be the only surviving copy of the certificate is numbered 124, so she did entice some investors with her Nolan Creek promise.[8]

It might be misleading to indicate the news stories that appeared about Nellie in the Juneau, Anchorage, or Fairbanks newspapers. Her life was unusual, and the occasional notice in the *San Francisco Chronicle* and the *New York Times* is an indication that her career was not commonplace.

In 1921, while "outside," she was featured in a full page article in *Sunset*, complete with photograph. In California she stated that she hoped to be appointed U.S. Deputy Marshal for the Koyukuk District. She figured she was as well qualified as any to hold the position:

> No, not the enforcement of the two-gun man. I wouldn't think of using force on anybody, particularly those boys up there. You see, they look on me as a sort of mother and they wouldn't think of doing anything wrong while I was round. I've been all through Alaska dozens of times but I've never been troubled by bad men. There isn't a man in Alaska who doesn't take off his hat whenever he meets me — and they always stop swearing when I come round, too. I wouldn't have any trouble in keeping order, because everybody's orderly when I'm round, anyway![9]

At the close of the interview Nellie showed some of her old spark, the longing for excitement. Although she loved Alaska, she was not tied to it. "I intend to have my claims worked, but after that, if anything turns up somewhere else, I'm likely to pull up stakes and start out." This was a seventy-six year old lady speaking.

The following winter Nellie went outside again seeking financing, this time to New York City. Nellie's trip from Nolan Creek to New York City was not exactly direct. She traveled the 750 miles from Nolan Creek by dogsled, overland, to Anchorage, in a gruelling trip which took seventeen days. Even in Alaska such travel bordered on madness. She was referred to as "the only white woman in Alaska who at her age, travels over the roughest country in any weather by dog team."[10] For a comparison, the distance from Nolan Creek to Anchorage is similar to the distance from San Francisco to Los Angeles, as the crow flies. But dog teams do not fly, so on winding ground, on hill and plain, the trip could be better compared to the distance between Seattle and Los Angeles.

She returned in April of 1923, and the following excerpts from a letter to her nephew Mike Cunningham in Arizona explain some of her recent moves:

> I arrived here on the 7th of April. I came over the mountains on a fast dogtrain and a native driver. The Northern Commercial Company had everything arranged for me. The sled only turned over once. I had a little roll in the snow. I didn't see the party that I wanted in New York he was in Florida. I will see you late in the fall. I will remain here five months. I left my clothes in Fairbanks. I came over the mountains very light. . . . I am feeling fine after my long trip. It is snowing here today. We don't expect the breakup not before the middle of May. (That is the spring breakup of ice on the creek). I am fixing up my cabin. I didn't move up from Wiseman. I will go up in a week's time then the cabin will be fixed and wood cut. My dear I've taken good care of myself I got off the river before the breakup came and believe me I got off of it in a hurry.[11]

Nellie Cashman, the seasoned, mature prospector-explorer-miner. (Sunset, May, 1921)

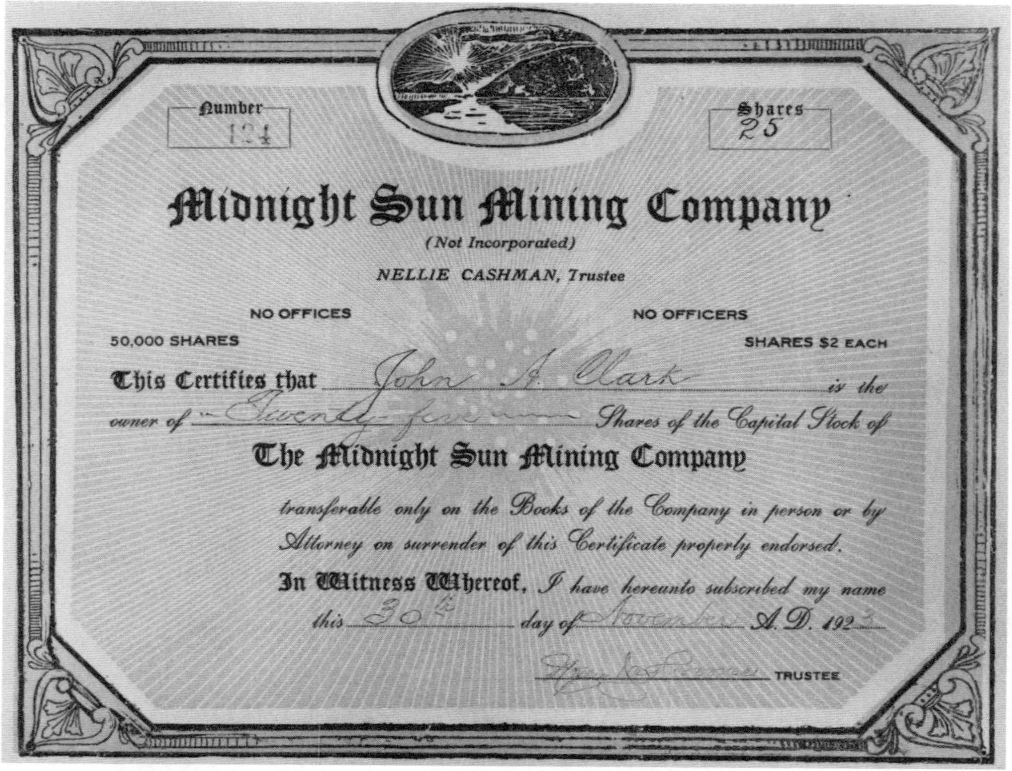
Number 124

Shares 25

Midnight Sun Mining Company

(Not Incorporated)

NELLIE CASHMAN, Trustee

NO OFFICES — NO OFFICERS

50,000 SHARES — SHARES $2 EACH

This Certifies that John A. Clark is the owner of Twenty-five Shares of the Capital Stock of

The Midnight Sun Mining Company

transferable only on the Books of the Company in person or by Attorney on surrender of this Certificate properly endorsed.

In Witness Whereof, I have hereunto subscribed my name this 30th day of November A.D. 1923

TRUSTEE

Shortly before her death, in order to raise funds to work her claims with hydraulic machinery, Nellie formed this company. (Rasmuson Library, University of Alaska-Fairbanks)

Her comment about Wiseman meant that Nellie was staying in that settlement instead of going a few miles farther up Wiseman Creek to Nolan Creek, where her claims were. She had cabins in both places, but Nolan Creek was her working center, while Wiseman was the social center of the entire Upper Koyukuk country.

The following winter Nellie went outside again, to visit relatives in Arizona and to try to obtain additional financing in New York. The New York trip did not materialize, so she wandered around Arizona for a few weeks, and was interviewed in Tucson, a city she had not visited in thirty-two years. The journalist compared her return to that of Rip Van Winkle.

Most of the interview was a rehashing of her career in Arizona, and of her current activities in Alaska. She did give a reason for leaving Arizona years ago: "when things began to be civilized, I left for Alaska, where I could pioneer again." There she had found men of "pure gold," and she loved her life among the "sturdy, uncultured, rough men."[12] On this trip to Tucson, Nellie also visited the newly consecrated Bishop Gerche of that diocese. She presented him with a check for $1,000, which she hoped would be spent on renovations of the church in Tombstone.[13]

Nellie Cashman, the hardy pioneer, the wanderer of western North America, got old, and her body broke down. She never did return to her favorite camp in North America, the placer ground at Nolan Creek. Upon leaving Arizona she became ill, kept going north. She stopped in Fairbanks and made arrangements for some mining equipment to be sent north, then proceeded to the Koyukuk River. She went as far as Bettles, within eighty miles of her claims, but could go no more.

Realizing that she had severe health problems, Nellie went down the river with Sam Dubin on his "Teddy H.," the common transport for years on the Koyukuk. When they reached the junction of the Yukon River they went south a few miles to Nulato, where Nellie stopped with the Sisters of St. Ann.[14]

The Sisters of St. Ann realized at once that Nellie was severely ill; they suspected pneumonia. There was no physician at Nulato, so it was decided that Sister Mary Claude would accompany Nellie to the hospital in Fairbanks. They left Nulato on the steamer "General Jacobs" on August 5 for Fairbanks. When Nellie boarded she noticed an elderly steward, Louis Iverson, a longtime acquaintance. She slapped him on the back with a "Hello, young fellow," and said she would soon be back, "even if I have to come back in an airplane." This was Nellie's birthday, so the captain arranged for a party with a special birthday cake in her honor.[15]

Upon reaching Fairbanks Nellie was taken to the St. Joseph's Hospital, another institution operated by the Sisters of St. Ann. Her condition was diagnosed as a combination of pneumonia and rheumatism, and it was decided that she needed further care at an institution in the states. Nellie was sent south, and she arrived in Seattle in the first week of October where she was installed in Providence Hospital.[16]

Within days another decision was made. Nellie wished to go to Victoria, to enter St. Joseph's Hospital, and she arrived there on October 9. Such are the roundabouts of history. This hospital, operated by the Sisters of St. Ann, was the same institution that received a huge impetus at the time of construction when Nellie Cashman passed the hat amongst the miners in the Cassiar District for gold dust. She had come home to a hospital she had helped build, which was operated by her friends of fifty years, the Sisters of St. Ann.

Nellie was in bad shape when she arrived, said she was "all in," and that "I am coming home to die." This surprised no one who saw her, as she was frail and fatigued, though still pleasant and ready to joke. She had many conversations with the Sisters, especially Sister Mary Mark, who was in charge. Her attending physician was Dr. Barrett, her old friend from St. Joseph's Hospital in Dawson. She was given the last rites of the church a few weeks after her arrival, as all agreed that Nellie would never recover.

She began to put her house in order, and arrangements were made to pay various bills through her banks in Victoria, Seattle,

The Sisters of St. Ann in Dawson, 1904, Nellie's last year there. She had close ties to this group from 1874, in British Columbia, and at subsequent missions in the Yukon and in Alaska. She died in their midst in Victoria in 1925. (Archives, Sisters of St. Ann, Victoria)

and Fairbanks. She dictated her will on November 7, giving most personal things to nephew Mike Cunningham and his family, and all clothing, other personal possessions, and what remained to Sister Mary Mark for whatever charitable purposes she preferred. As executor she named R. E. McIntyre of Nolan Creek, "Bob" with the screechy voice, with whom she had worked claims for twenty years.[17]

On January 4, 1925, Nellie was "visited by Alaska Sisters who knew her sterling worth." She died in their company. Dr. W. T. Barrett and Sister Mary Mark decided on a post mortem examination in order to determine the cause of death, "as they both felt that some day Nellie's life would be written." Dr. Barrett concluded that she died of "unresolved pneumonia." Three days later, following funeral ceremonies in St. Andrew's Catholic Cathedral, she was buried in the cemetery at Ross Bay, Victoria, adjacent to the plot of the Sisters of St. Ann.[18]

Although Nellie had been in the distant northland for over twenty-five years, she had earned a reputation as a "character," and word of her death was widespread. Rich prose in her obituaries could of course be seen in newspapers of Dawson, Fairbanks, Tucson, Tombstone, and Bisbee. Yet, the fact that important newspapers such as the *New York Times, Los Angeles Times, San Francisco Chronicle,* and *Victoria Daily Colonist* gave space to the passing of this prospector-miner reveals that people had an interest in this elderly lady who was stopped only by death. As one writer put it, "Pioneer Woman Prospector Goes on Last Stampede."[19]

The fact that she was more than a "character" is borne out by the notice she was given by the mining profession. The *Engineering and Mining Journal-Press* acknowledged her death as the passing of a colleague, stating that she was one of the first women to enter the Alaska gold fields as a prospector, that she had grubstaked and accompanied many expeditions, and "was held in high regard by a very wide circle of acquaintances."[20]

One question that comes immediately to mind is to ask how successful Nellie Cashman had been with her years of mining in the Upper Koyukuk. Her will is specific but deals only with jewelry, clothes, and other personal possessions. She did have bank accounts in three or four cities, enough to pay hundreds of dollars of bills during her last month. However, she does not mention these accounts in her will. It may be that all Bob McIntyre had to "execute" was Nellie's claims and equipment on Nolan Creek.

Another possibility is that Nellie did her banking in conjunction with her nephew Mike Cunningham. Mike had started his career as a cashier in Arizona in 1899, and by the 1920s was president of banks in Bisbee and Douglas, Cochise County. Mike was Nellie's favorite, and he also owned a few of the Nolan Creek claims with her. There is also a hint that some of the investors in the Midnight Sun Mining Company wondered about their money after Nellie's death, and that Mike personally reimbursed them. He could have, and he probably did so.[21]

Nellie Cashman was no fool, and she lived for twenty years north of the Arctic Circle, in some of the most difficult living conditions on earth. She most likely earned enough from her mining to have a comfortable life, and she certainly kept up her faith in her claims. Nellie was extravagant, spending many thousands of dollars on equipment, not only in Fairbanks but also in Seattle and San Francisco. Furthermore, in her last ten years, Nellie took at least annual trips to Fairbanks, and at least seven trips to points like Tucson, Seattle, St. Louis, Minneapolis, and New York City. This says something for the productivity of her Nolan Creek claims.

Much of Nellie's money went to causes, and when she was down and out but there was still a need for hospitals or churches, Nellie did not hesitate to ask others to contribute. In flush times, there was no brake on Nellie's spending. The following example is from an account of Sister Mary Theodore, who wrote of Nellie in the Yukon, Alaska, and British Columbia:

> She was not only charitable, she was charity personified. She gave and gave, both from impulse and from motives of purest Christianity. . . .In times of sickness, of epidemics,

> she was present with helping hand. The time was drawing near for Sister Mary Pudentienne to pronounce her final vows and receive the ring of her profession; but, in that land teeming with gold, there was no silver for the tiny ring. Sister does not know how Nellie contrived to do it, but the ring came, "and," Sister remarked, "it must have cost a penny." Roses, valued at one hundred dollars, came for the chapel, — ordered from a distance of some two thousand miles. Sometimes a cart laden with pillows and comforters would be delivered at the hospital in her name. What she gave to the hospital was only a side outlet to her donations. She contributed largely to the Church; to the reckless needy when she found them out, or when they found her.[22]

This was not an aberration but part of Nellie Cashman's lifelong plan of helping when she could. There are dozens of documented examples of her charities, but probably actually hundreds that she performed. Some of her major efforts were important to the people and the communities in which she lived, such as the hospital in Victoria, the church and hospital in Tombstone, and the church in Dawson. Also, scattered throughout the northland, where the Sisters of St. Ann had hospitals, schools, orphanages, and missions, Nellie Cashman was known as a steady financial and spiritual contributor.

Much of Nellie's life became consumed by her practice of helping and giving. Legends grew about her, many of them bearing little relation to the truth. There were too many stories of the "Angel of Tombstone," "The Miner's Angel," "Nellie Cashman, Savior in the Wilderness," and so forth.

Because of her staunch Catholicism and her lifelong practice of giving, journalists, editors, even her friends tended to forget or placed aside her most important characteristic, that of a stampeder.

Nellie Cashman thrived in mining camps, but they had to be active camps, or at least camps with promise of a bonanza future. She had no loyalty to buildings or locations. She needed the vitality of a vigorous, young mining community. Of course, most of the people who were there shared very few of her interests. Precious metals camps favor the young, the uneducated, the crude, the boomers. All of them, including Nellie Cashman, were looking for something to happen quickly, and if it didn't, they moved on.

Most stampeders, including Nellie Cashman, wanted instant wealth. However, what one did or planned to do with the wealth could certainly differ. Many who stampeded to the California gold fields, to the silver camps of Arizona and Nevada, or to the golden sands of Nome could have been dreaming of owning a cotton plantation in Alabama, running a saloon in Toledo, or operating a meat market in Dubuque.

There was never any mystery or question about Nellie's search for the Big Bonanza. She wanted to find it in order to share it. The evidence for this is in her long life: when she had money, she gave it away.

Some may question the Nolan Creek years. Her good production there seems to have been in the period 1907-12. Why, then, did she remain until her death in 1925? This does not appear to fit in with the "stampeder" mentality.

In many ways, it actually does fit. Nolan Creek for Nellie was a continual provider of gold, although not always plentiful. Yet, she always believed that with proper financing, and better, larger hydraulic equipment, she could penetrate to bedrock, drift out, and wash tons of gravel and hit the big bonanza. And when she did, she would give it all away. The following is a diary entry of January 9, 1925, from Nulato, where the Sisters of St. Ann had just learned of Nellie's death:

> Reported Miss Nellie Cashman died at St. Ann's Hospital, Victoria. She is a great pioneer, always above reproach, very charitable, remarkably devout to the Holy Souls, in whose behalf she meant to make a foundation with perpetual suffrages; the great object of her mining on the Upper Koyukuk river.[23]

There is no reason to doubt these plans, as Nellie's entire life seemed aimed at such a cause.

There is an obvious fascination with the fact that Nellie was a woman in a man's profession. Not only that, but under the frontier conditions in which she explored, prospected, and mined, she had placed herself in a society in which few decent women could be found.

Fred Lockley, who had known Nellie in Alaska, interviewed her in Portland a few years before she died, and she stated that in the wilderness, there was an unwritten pact between her and the men:

> Every man I met up north was my protector, and any man I ever met, if he needed my help, got it, whether it was a hot meal, nursing, mothering or whatever else he needed. After all, we pass this way only once, and it's up to us to help our fellows when they need our help.[24]

Nellie's hope for Nolan Creek turned into reality. Even though the bonanza days had passed, new developments in 1925 and 1926 led to a substantial increase in gold production. A new company had been organized, the various claims in the region consolidated, and plans were underway to build a ditch forty-five miles long, to bring water to the newly thriving placer benches on the Hammond River and Nolan Creek. Nellie had been proven correct again, but died before the news reached her.[25]

NOTES TO THE CHAPTERS

Chapter One
Getting Started

1. Edward MacLysaght, *More Irish Families* (New York: Barnes & Noble, Inc., 1960), p. 158.

2. The marriage and birth records from Midleton Parish are due to the diligent searches of Tim Cadogan of the Cork County Library and Canon B. Troy of Midleton Parish. Sister Mary Loyola gathered considerable correspondence and documentation on Nellie in the 1970s, which are now organized as the Cashman Papers in the Archives, Sisters of St. Ann, Victoria, British Columbia; hereinafter referred to as The Cashman Papers. Provincial death records, and the tombstone in the cemetery at Ross Bay, Victoria, indicate a birth date of 1844, which is much closer than most of the other conjectures about the year of her birth.

3. A sound, detailed study, with statistics and graphs, is Kerby A. Miller, *Emigrants and Exiles: Ireland and the Irish Exodus to North America* (New York: Oxford University Press, 1985).

4. *Ibid.*, p. 364.

5. From an interview with Fred Lockley, who had talked and visited with Nellie on several occasions; in the Seattle *Alaska Weekly*, February 19, 1932.

6. See Miller, *Emigrants and Exiles*, pp. 406-08. Some of this is also drawn from Laurie K. Mercier, "We Are Women Irish," *Montana Magazine of History*, XLIV (Winter, 1994), 29-41. See also Janet Nolan, *Ourselves Alone: Women's Emigration from Ireland, 1885-1920* (Lexington: University Press of Kentucky, 1989).

7. Frances Cashman appears in the Boston city directories for 1865 and 1866. Because directory information is often collected a year in advance, her listing in the 1866 directory does not conflict with the statement that the family went to California in 1865. The fact that Nellie operated a "lift in a public building" is not in much dispute, as she reportedly said so in a dozen interviews or more. However, some writers have romanticized her role, having her operate a lift in Washington, D.C., during the Civil War, where she served as a "Federal Spy," etc.

8. Lockley interview, *Alaska Weekly*, February 19, 1932.

9. See the opening pages of R. A. Burchell, *The San Francisco Irish, 1848-1880* (Berkeley: University of Calif. Press, 1980).

10. R. Hal Williams, *The Democratic Party and California Politics, 1880-1896* (Stanford: Stanford University Press, 1973), pp. 23-24.

11. *Langley's 1869 San Francisco Directory* on p. 180 shows Cunningham as pres. of the U.W.M. Boot & Shoe Co., at 336 Fifth St.; p. 142, Miss Ellen Cashman [obviously, Nellie] is also listed as living as 336 Fifth St.

12. Cunningham's boot & shoe ventures appear in each San Francisco directory through 1881, along with his various changes of address. The *1872 Great Register, San Francisco County,* on p. 85, contains details of his arrival in this country and of his naturalization.

13. *Mining & Scientific Press,* January 11, 1873.

14. The data for buying and selling the Pioche boarding house are in a letter from Sister Mary Loyola to Mr. F. C. Brophy, May 22, 1973; Cashman Papers, Victoria. An advertisement for the "Miner's Boarding House" operated by Fanny Cashman appears in the *Ely Record,* September 4, 1872.

15. James W. Hulse, *Lincoln County, Nevada: 1864-1909* (Reno: University of Nevada Press, 1971), p. 27.

16. Such a strong conclusion is based on my research for *Virgil Earp: Western Peace Officer* (Encampment, Wy.: Affiliated Writers of America, 1994). Virgil served as Police Chief of Tombstone during its most troubled period, but gunfights and street violence were relatively rare in that community.

17. *Pioche Daily Record,* October 9, 1872.

18. *Ibid.,* November 21, 1872; March 2, 1873.

19. Dan Ashbaugh, *Nevada's Turbulent Yesterday* (Los Angeles: Westernlore Press, 1963), Pioche summary on pp. 21-37.

20. *Mining & Scientific Press,* September 6, 1873.

21. Fr. Scanlan was the Catholic priest, who in 1873 was replaced by Fr. D. Monteverde. The schools, churches, and other social organizations received frequent mention in the press. For example, the *Pioche Daily Record* of January 29, 1873, had a full column article on the hospital, listing each patient, cause of illness, and so forth. The issue of September 20, 1872, discussed Fr. Scanlan's work with the Benevolent Society of St. Vincent de Paul. The issue of March 18, 1873, devoted an entire column to the St. Patrick's Day activities.

22. *Pioche Daily Record,* October 4 & 15, 1873.

23. *Pioche Daily Record,* September 20, 1872.

24. A news note about Nellie, and her first ad, appeared in the *Pioche Daily Record* of April 15, 1873.

25. Hulse, *Lincoln County, Nevada,* p. 22. A good, near-contemporary appraisal of the new, wild community of Pioche is in H. H. Bancroft, *History of Nevada, Colorado, and Wyoming* (San Francisco: History Company, 1890), pp. 272-73.

26. The *Pioche Daily Record* of August 24, 1873, mentioned the first party that had left for the mines in Inyo County; this would soon turn into a semi-stampede.

27. *Mining & Scientific Press*, September 6, 1873, a detailed three-column article on mining, mills, and public facilities in Pioche and district.

28. Sister Mary Loyola to Brophy, May 22, 1973, Cashman Papers, Victoria. In the *Pioche Daily Record* of December 24, 1874, Jacobs & Sultan was listed as in "Wholesale and Retail."

29. *Pioche Daily Record*, October 22, 23, 1873.

Chapter Two
Canadian Challenge

1. A typical work of this type is Will H. Chase, *Pioneers of Alaska: The Trail Blazers of Bygone Days* (Kansas City: Burton Pub., 1951). In discussing the pioneers on p. 14, Chase points out that the first activities in the North of consequence were the Cassiar discoveries in British Columbia, beginning in 1861.

2. I have an extensive file on "Buck" Choquette, based on Hudson's Bay Co. Archives, the Victoria newspapers, and many geological and mining reports. The Hudson's Bay officials had a real problem with Choquette. He was the best trader by far, yet he tended to put his own interests ahead of the Company. They reluctantly hired him on several occasions.

3. Choquette had the "fever," and lost it only when death took him in 1898, in the Yukon gold camp of Dawson. Throughout the 1860s, Choquette continued to prospect and find some gold up the Stikine, and there are a dozen or so accounts about him in the Victoria newspapers. In January of 1875 he was listed in the advertisements as a leading "Indian Interpreter & Explorer, Dealer in Groceries, Provisions, & Miners' Outfits" in the Cassiar District; *Daily British Colonist*, January 28, 1875.

4. Two standard works that summarize Thibert's discovery are R. M. Patterson, *Trail to the Interior* (New York: Wm. Morrow & Co., 1966), and Clarence L. Andrews, *Wrangell and the Gold of the Cassiar* (Seattle: Luke Tinker, 1937).

5. I do not know what happened to Thibert, but he certainly caught his share of gold "fever." He remained in the area, prospecting and mining years after the rush. In 1886-87, when Canadian geologist George Dawson studied the area, one of his best informants was Thibert; George M. Dawson, *Report on an Exploration in the Yukon District, N.W.T., and Adjacent Northern Portion of British Columbia, 1887* (Montreal: Dawson Bros., 1888), p. 111.

6. For Stikine meetings in Nevada, see *Pioche Daily Record*, March 16, March 21, 1873.

7. *Alaska Weekly*, February 19, 1932. In an earlier interview, while on her way to the Klondike in 1898, Nellie reminisced about going to Cassiar and stated she went north "with a party of 200 miners from Nevada." There actually were many dozens of Nevada miners up in the Cassiar, but the "200" probably refers to the party from various places that she joined in San Francisco enroute to Victoria; *Daily British Colonist*, February 15, 1898.

8. The 2" x 4" photo, with studio address on the reverse, is in the Cashman Papers. This photo was the basis for the celebrated oil painting of Nellie, done in China through the arrangements of her Chinese cook in Tombstone.

9. Nellie provided most of these details in an interview published in the *Daily British Colonist,* February 15, 1898.

10. Other detailed accounts of the Stikine country are in the issues of January 31, April 18, and June 20, 1874.

11. John Y. Smith of Victoria was in the Cassiar during 1872-74 and later wrote a detailed account of the geography, costs, hardships, transportation, and successes and failures of the Dease Lake area; "Experience of a Greenhand in a Gold Mining Camp," typescript, Cashman Papers.

12. The *Daily British Colonist* carried frequent reports and accounts of the Cassiar diggings. The issues of August 23, August 28, and August 30 are particularly important, as these reported on what turned out to be the height of the rush.

13. *Daily British Colonist,* February 15, 1898.

14. *Daily British Colonist,* February 5, 1875.

15. *Ibid.,* March 6, 1875.

16. Nellie's comments on the rescue expedition can be seen in the *Colonist* for February 15, 1898, and in the *Alaska Weekly* of February 19, 1932, based on information she had given Fred Lockley.

17. The money sent to her mother is mentioned in the Barkerville *Cariboo Sentinel,* July 17, 1875. The incident of the St. Joseph's Hospital fund is contained in several memos and letters in the Cashman Papers, Sisters of St. Ann, Victoria.

18. The geologist Josiah E. Spurr wrote an article on the Yukon Gold District, with introductory comments on the Cassiar, in the *Eighteenth Annual Report of the United States Geological Survey* (Washington: GPO, 1898), Part III. His production chart is on p. 114. The reason for inclusion was explained on p. 112: "The Cassiar district was, in a way, the training school for Yukon miners." Another useful summary of the district is in H. H. Bancroft, *History of British Columbia, 1792-1887* (San Francisco: History Co., 1887), pp. 559-64.

19. *Daily British Colonist,* June 16, 1876.

Chapter 3
Arizona Calls

1. The railroad history is well summarized in H. H. Bancroft, *History of Arizona and New Mexico* (San Francisco: History Company, 1889), p. 605.

2. C. L. Sonnichsen, *Tucson: Life and Times of an American City* (Norman: University of Oklahoma Press, 1982), p. 105.

3. In Nellie's interview with Bernice Cosulich, *Arizona Daily Star,* January 18, 1924.

4. Either interviewer Cosulich or Nellie became confused, as she sup-

posedly claimed to have taken the stage from Los Angeles to Yuma, whereas she clearly had taken the train; the stage route from Yuma to Tucson paralleled the newly laid rail lines for some miles.

5. *Daily Arizona Citizen,* July 29, 1879. The *Arizona Daily Star* of January 19, 1879, reported that Miss Jennie Swift of Los Angeles arrived in Tucson "with a view of starting a private school." She may have become a partner with Nellie in the Delmonico. When Nellie later moved to Tombstone, Jennie Swift followed and was a business partner for a few months.

6. John Clum, "Nellie Cashman," *Arizona Historical Review,* III (January, 1931), 10.

7. Adolphus Henry Noon, "Arizona," *Chicago Tribune,* December 5, 1879; Noon wrote a series of articles on his Southwestern trip.

8. Sonnichsen, *Tucson,* pp. 105-12.

9. Lonnie E. Underhill, "The Tombstone Discovery: Recollections of Ed Schieffelin & Richard Gird," *Arizona and the West,* XXI (Spring, 1979), 37-75.

10. I dealt extensively with Tombstone in my *Virgil Earp* and *The Earp Papers: In a Brother's Image* (Encampment, Wy.: Affiliated Writers of America, 1994).

11. There is, of course, a rich geological literature about the Tombstone ore and geology. I prefer the contemporary accounts; William P. Blake, "The Geology and Veins of Tombstone, Arizona," *Transactions, American Institute of Mining Engineers,* X (1882), 334-45, and the many articles in the 1879-81 issues of the San Francisco *Mining & Scientific Press.*

12. An excellent contemporary history of the town's founding, with illustrations, can be found in *Arizona Quarterly Illustrated,* July, 1880; copy in Huntington Library, San Marino, California.

13. The business directory was in the *Tombstone Weekly Nugget* of June 10, 1880; a typical ad for the Nevada Boot & Shoe store was in the *Nugget* of May 20, 1880.

14. Advertisements were in the *Nugget* of June 24 and the *Epitaph* of June 19. The *Star* comment is in the issue of June 13, 1880.

15. *Nugget,* June 24, 1880.

16. The *Epitaph* had articles on July 7 and July 15, 1882, with details of the territorial census which was based on the 1881 residency. Other towns in Cochise County with substantial population were Benson, 820; Contention 452; Charleston 423; Bisbee, 410, and Willcox, 324.

17. Advertisements in the *Nugget* of July 29 and October 10, 1880, and the *Epitaph* of October 24 and 29, 1880. Every issue of these newspapers carried advertisements or notices of Nellie's ventures.

18. Tombstone postal records for this era indicate about eighty money orders to various merchants, some listed here. She also would frequently send $50 money orders to Kate O'Hara. Kate had been involved in the early stages of the Tombstone cash store, before Nellie got involved with groceries, so Kate probably concentrated on purchase of clothing in San Francisco; Cashman Papers.

19. For example, the *Nugget* of November 3 ran an ad for the Nevada Boot & Shoe Store. It is not clear if the Arcade closed its doors temporarily;

the *Epitaph* of August 28, 1880, still carried an Arcade ad in Pascholy's name. In the issue of September 18, he asked all who owed accounts at the Arcade to "come forward and settle."

20. Deeds, Real Estate, Cochise County, Book 1, pp. 464-65, County Courthouse, Bisbee.

21. Nellie's work was summarized in a letter of Fr. Thomas Doyle, Sacred Heart Church, Tombstone, to Sister Mary Loyola, July 7, 1971, Cashman Papers.

22. It is difficult to separate the church from the hospital fund drives, as the newspapers frequently commented on both activities. For the Catholic church in Tombstone, and the origins of the County Hospital, see the following for useful comments and data: *Nugget*, October 19, 1880; *Epitaph*, September 21, 23, 26, 28, October 3, 5, 7, November 5, 27, 28. The November 27 article is particularly relevant, because it is the announcement of the first Catholic services, to be held the following day. The last sentence reads: "Miss Nellie Cashman, to whose zealous, active work much of this success has been due, desires us to thank the public for the generous manner in which they responded to her calls for aid in this worthy cause."

23. *Nugget* of June 9 has an advertisement for the ball, with a complete list of all committee members.

24. Details are in a notice in the *Epitaph* of June 5, 1881.

25. Korka ran a notice in the *Epitaph* of June 9, stating that he would keep the same hours, and that he was "Successor to Miss Nellie Cashman."

26. Nellie's plans were outlined in the *Epitaph* of July 8, 1881. For other interesting comments on Bisbee in these months, see *Epitaph*, March 11, March 15, April 17, 1881.

27. The *Los Angeles Herald* of June 24 listed the destroyed structures and the amount of insurance on the buildings. Included were Arcade Restaurant, $2,000; Arcade cigar store, $2,000; Arcade Saloon, $1,500.

28. *Nugget*, July 10, 1881.

29. Mentioned in the *Nugget* of October 14, 1881.

30. Quite a few money orders to Nellie's mother are listed in the Cashman Papers. These papers also contain many birth, marriage, and death records of the Cashman-Cunningham families.

31. The lease announcement and building repairs are mentioned in the *Epitaph* of October 2 and in the *Nugget* of November 10. The Delmonico Lodging House is also listed in *McKenney's Business Directory. . .*1882-3, on p. 284.

32. Articles in the *Nugget* of October 4 and the *Epitaph* of October 4 carried flattering articles on the new operation.

33. Practically any issue of the Tombstone newspapers of this era contained advertisement, notices, or news notes that included mention of the Russ House. The Russ House was built in December of 1880, and Chas. D. Stuart was the manager when it opened that month; *Nugget*, December 17, 1880.

34. See also the issue of October 20, 1881: "Her soul goes out in sympathy wherever there's human suffering," and the *Nugget* of October 30.

35. *Epitaph*, November 30, 1881.

36. See the *Epitaph* of December 17 for Nellie, and the *Nugget* of December 31 for the Delmonico.

37. *Epitaph*, March 20, 1882. The Sisters were said to be "of St. Louis," but that was the Mother House; the Sisters of St. Joseph operated schools and a hospital Tucson.

38. John Pleasant Gray Reminiscences, pp. 39-40, Library, Arizona Historical Society.

39. *Epitaph*, April 20, 1882.

40. There would be no account of the fire from the *Nugget*, as that feisty organ would live no more. The above paragraphs are from the detailed account which appeared in the *Epitaph* of May 27, 1882.

Chapter Four
Weak Legends, Great Adventures

1. Her nephew Michael Cunningham had a successful career as a Bisbee banker, and there is considerable detail on his background, his education, and his closeness to Aunt Nellie. See, for example, *History of Arizona: Biographical* (Phoenix: Record Pub., 1930), III, 87. The various members of the Cashman-Cunningham family were buried in Calvary Cemetery, but later all burials there were transferred to the Holy Cross Cemetery, Colma, near San Francisco. Sister Mary Loyola obtained many vital documents on the Cunningham family, including the hard-to-follow story of Fanny's death and burial in San Francisco; these are in the Cashman Papers.

2. I have done much research and field work in this area. For a useful general work which I co-authored with William Mason and David Zarate, see *The Mines of Baja California Norte* (Los Angeles: Natural History Museum, 1992).

3. Francis J. Weber, *The Peninsular California Missions, 1808-1880* (Los Angeles: The Author, 1979), p. 41-43.

4. The earliest account I have spotted is in the *San Diego Union* of April 15, 1883, and it is mentioned that Senor Don Jose Gradias had brought to the city over $1,000 worth of gold from the diggings.

5. See, for example, the route description in *Arizona Daily Star*, May 17, 1883.

6. Quote from the *San Diego Union* of June 2, 1883. For another comment on the San Quentin route, see the same paper of May 31, 1883; Capt. Gregory of the "Catura" would take as many as 25 passengers on his schooner for $25 each, and they were limited to 200 pounds of baggage each.

7. Figures published in Hermosillo *La Constitucion*, June 1, 1883; Hermosillo was the capital of the State of Sonora.

8. Reprinted in the *San Diego Union* of May 25, 1882, from an issue of May 22.

9. John Anderson wrote a letter from Guaymas on May 17 describing Riveroll's arrival, and this was printed in the *Arizona Daily Star* of May 20, 1883.

10. There is even a literature on Nellie's Baja California adventure, and it is so unbelievably bad that I hesitate to mention it. Not only are most details wrong, but the accounts don't even have the right year, pointing to 1884 instead of 1883. A few of these weird accounts are Clum, *Arizona Historical Review,* 1931, pp. 23-25; Louise Cheney Auer, "Nellie Cashman and Her Lost Gold," *Desert Magazine,* XXVI (October, 1963), 24-26; Charles Hogan, "Did Nellie Cashman Find Gold?" *The West,* May, 1965, p. 17; Choral Pepper, "An Angel and the Valley Gold," *Desert Magazine,* XXXVIII (December, 1975), 24-27.

11. The *Arizona Daily Star* of May 16, 1883, devoted two columns to the excursion, praising Guaymas as "the coming metropolitan port of the gulf."

12. Quote is from the San Diego *Sun* of June 15, 1883. Other details of planning and departure are in the *Arizona Daily Star* of May 23 and 24.

13. Willard, the U.S. Consul in Guaymas, supplied many details to the press; he mentioned that the Cashman party had chartered a vessel and had purchased burros; *San Diego Union,* June 9, 1883.

14. The best detailed account of Nellie's disaster is in the *San Francisco Chronicle* of June 12, 1883. Important comments are in John Hart's diary, *Arizona Weekly Citizen,* June 16, 1883.

15. Good details of the topography and climate are in *Arizona Daily Star* of June 8 and June 12, 1883.

16. *Arizona Weekly Citizen,* June 16, 1883, Hart's diary. Other comments on the near disaster are from the *Arizona Daily Star* of June 12, 1883, and *San Francisco Chronicle,* June 12, 1883. One account, in the *Weekly Arizona Citizen* of June 16, 1883, has this line: "Especial mention is due Pete Gill and his boys for their assistance, under these circumstances." Who Gill was, or what he did in aid of the party, is not known.

17. *Ibid.*

18. *San Diego Union,* June 13, 1883. Another example urging caution before proceeding to the gold fields was written by Dr. George R. Ghiselin and appeared in the San Diego *Sun* of June 8, 1883. The *Los Angeles Times* of June 5 had two long articles warning of the dangers of going to these gold fields; one article concerned "The Cold Facts about the Placer Gold Mines of Lower California."

19. *Weekly Arizona Citizen,* June 15, 1883.

20. *Los Angeles Times,* July 4, 1883, news note from a Tucson newspaper of June 30.

21. Most of the mining claim location notices, sales, and other mining activities were brought to my attention by Cochise County Recorder Christine Rhodes. The specific details in the above paragraphs may be found in Record of Mines, Books 1, 8 11; Deeds of Mines, Books 5,7,8,9, and 11.

22. A recent article on this affair is George M. Ellis, "Sheriff Jerome L. Ward and the Bisbee Massacre of 1883," *Journal of Arizona History,* XXXV (Autumn, 1994), 315-42; Dr. Goodfellow quote is on p. 332. Of the newspaper coverage I followed primarily the detailed reporting of the Tucson *Arizona Daily Star,* issues of March 27, 28, 29, 1884.

23. *Arizona Daily Star*, March 27, 1884.

24. Most of the accounts dealing with Nellie seem to be based on Clum's narrative; *Arizona Historical Review*, 1931, pp. 16-20. He even has Nellie acting as Mother Confessor, without explaining why this was needed with at least three clergymen present.

25. *Arizona Daily Star*, March 29, 1884.

26. Issue of April 4, 1884.

27. John Pleasant Grey Reminiscences, pp. 39-40, Library, Arizona Historical Society.

28. Clum, *Arizona Historical Review*, 1931, pp. 20-21; Fred Dodge-John Clum letters of the late 1920s, Dodge Collection, Huntington Library.

29. *Arizona Weekly Citizen*, August 9, 1884.

30. *Ibid.*, August 16, 1884, copied from the Tombstone *Epitaph*.

31. *Ibid.*; see also a detailed account in the *Arizona Daily Star*, August 10, 1884.

32. This was Company C, led by Captain Wm. N. Tilsdall. They reported to Sheriff Ward at the courthouse, then proceeded to Contention Hill, where they camped for several days; *Arizona Daily Star*, August 14, 1884.

33. Quote from *Arizona Weekly Citizen*, August 16, 1884.

34. On August 10 the *Epitaph* had published a notice calling on the men to disperse and to turn over the leaders of the mob to the authorities. Leach, other Grand Central officers, and Attorneys Walker and Fell signed the notice; Gage did not, and he was not mentioned. I suspect that Gage was in Tucson on business and that Leach was temporarily in charge.

35. From 1900 to 1924 Frank Cullen Brophy saw Nellie several times, as she visited the Cunninghams and Brophys in Bisbee on her trips out of Alaska. However, the material Brophy wrote about Nellie contains the same pitfalls, the same absurd errors of geography and chronology as the other sources on Nellie; see, for example, Frank Cullen Brophy, *Arizona Sketch Book* (Phoenix: Ampco Press, 1952), pp. 72-81, "Frontier Angel." Another Brophy book is *Though Far Away* (Glendale, CA: A. H. Clark, 1940). Clum, although ex-mayor of Tombstone, was gone from the city during the years of these events, and he relied on Mike Cunningham for his "facts."

36. The sale and inventory, dated June 13, 1884, is in Cochise County, Bill of Sale, Book 1, 188-92.

37. The incident is covered in some detail in Suzann Ledbetter, *Nellie Cashman: Prospector and Trailblazer* (El Paso: Texas Western Press, 1993), pp. 35-36.

38. Tombstone *Record-Epitaph*, November 4, 1885.

39. *Ibid.*, November 12, 1885.

40. Cochise County, Deed of Mines, Book 9, 161, dated November 9, 1885.

41. The Tombstone *Epitaph* of January 7, 1886, and the *Arizona Daily Star* of January 6 and January 10 featured news notes about Mrs. Jones and Nellie.

42. Advertisement in Tombstone *Epitaph*, February 3, 1886. In the 1880

California Census, Thomas was listed as age eight, meaning a birth year of 1871 or 1872.

43. The quotes are from an *Epitaph* story of January 31, 1886; the issue of October 20, 1886, carried a story headlined "Noble Nogales."

44. Nellie comments in the *Arizona Daily Star* of June 29, July 27, 1886.

Chapter Five
Nellie Buries Tombstone

1. For Miners Dream claim, Cochise County, Record of Mines, Book 11, 237, dated January 1, 1887. James E. & Barbara H. Sherman, *Ghost Towns and Mining Camps of New Mexico* (Norman: Univ. of Oklahoma Press, 1974), pp. 124-28. Kingston had a postoffice from 1882 through 1957.

2. James A. McKenna, *Black Range Tales* (New York: Wilson-Erickson, Inc., 1936), pp. 142-43. McKenna was in the Black Range District throughout the 1880s, and his account is humorous but based on first-hand information.

3. *Ibid.*, pp. 149-50.

4. I have seen a dozen or so references to Doheney as Nellie's dishwasher, but nothing from that era. For example, the *Tucson Daily Citizen* of April 5, 1965, quotes John Gilchriese reporting that Nellie considered Doheney "one of the best pearl divers I had."

5. McKenna, *Black Range Tales*, p. 119.

6. Later, in the Yukon, a correspondent recalled first meeting Nellie in Kingston, where "Miss Cashman at that time was keeping the Cashman house, the leading hotel of the district"; *Klondike Nugget*, April 21, 1900.

7. Some of the early accounts are in *Arizona Daily Star*, December 13, 14, 17, and 20, 1888; *Phoenix Daily Herald*, December 12, 22, 31, 1888.

8. *Tombstone Prospector*, December 23, 1888.

9. For Shibell, *Arizona Daily Star*, January 1, 1889; for Earp, *Arizona Daily Star*, February 12, 1889.

10. *Tombstone Prospector*, issues of January 5, 14, 15 mention her Phoenix trips.

11. Slightly different accounts appeared in *Phoenix Daily Herald*, January 28, 1889, and *Arizona Daily Star*, January 31, 1889.

12. *Phoenix Daily Herald*, February 25, 1889, carried a summary of his interview with the *Star*.

13. *Weekly Phoenix Herald*, August 1, 1889.

14. Issue of March 6, 1889.

15. Although the Harqua Hala was an important gold producer in western Arizona Territory, I know of no solid history of the district. I have done considerable research on the property, especially in the Guildhall Library, London, which has the company annual reports as well as prospectuses. The London *Mining Journal* for the 1890s has considerable information on the Harqua Hala.

16. *Arizona Gazette*, April 20, 1889.

17. These quotes and comments about the Quinns are based on stories of the Yuma *Arizona Sentinel*, February 23, March 23, March 30, 1889.

18. Item repeated in the *Tombstone Prospector* of November 25.

19. For Miners Dream, see Cochise County, Power of Attorney, Book 2, 247; Deed of Mines, Book 11, 107, 109, 136; Probate, Book 1, 85; these are mining and probate events from March through July 1, 1889. Historians at the National Museum in Bloemfontein and the Kimberly Public Library have not uncovered any evidence that Nellie was in South Africa; letter, Mrs. C. Duming, Kimberly Public Library, February 10, 1995.

20. *Phoenix Daily Herald*, December 6, 1889.

21. *Tombstone Prospector*, January 25, 1890.

22. *Arizona Daily Star*, January 24, 1890, article entitled "Nellie in Prescott."

23. The *Arizona Weekly Journal-Miner* of Prescott had a constant stream of positive articles in December, 1889, and January, 1890. For example, on December 18, 1889, it stated that "the mining interests of this section never look more favorable than at the present time."

24. Sister Mary Loyola in the early 1970s worked diligently on a sound chronology, but was also puzzled by the years 1890-95; Cashman Papers.

25. Interview with Mike Cunningham, *Bisbee Daily Review*, April 1, 1948.

26. *Tombstone Prospector*, October 30, 1894; January 7, 1895. Apparently a Prescott newspaper had criticized Nellie, as the *Prospector* of January 7 stated "the insinuation on the part of the Courier that Miss Nellie could not stomach her own vitals is a vile slander on one of our old favorites."

27. News clipping dated January 21, 1895, Sharlot Hall Museum, Prescott. The Elite had been first opened in 1890 by a Mr. Karns.

28. Victoria *Daily British Colonist*, February 15, 1898; this was an interview with Nellie when she reminisced about the trip to Juneau.

29. The Ensenada, Mexico, *Lower Californian* of February 7, 1896, described the Minas Prietas operation as consisting of ninety stamps working in three mills, with other mills planned. For Nellie in Minas Prietas see *Phoenix Weekly Herald*, June 13, 1895.

30. For Cripple Creek; a news clipping, Sharlot Hall Museum, Prescott, of October 8, 1895, states that Nellie had just arrived in Prescott from Cripple Creek.

31. Issue of October 24, 1895.

32. Issue of November 6, 1895, II, 14.

33. Issues of October 19, 1895, and February 1, 15, 27, and March 5 carried notices or ads for the Buffalo Hotel, or mentioned Nellie.

34. *Arizona Daily Star*, June 20, 1895.

35. Yuma *Arizona Sentinel*, July 17, 1897.

36. *Ibid.*, August 7, 1897, advertisement.

37. *Ibid.*, July 24, 1897, news note.

38. Wedding notice, *ibid.*, November 13, 1897; the bride was identified as

niece of Nellie, "the popular landlady of the Hotal Cashman." No issues of the newspaper are available for the remainder of 1897, the period in which Nellie disposed of the Hotel Cashman.

Chapter Six
From Desert Sand to Glacier Peaks

1. Lonnie E. Underhill, "The Tombstone Discovery: Recollections of Ed Schieffelin & Richard Gird," *Arizona and the West,* XXI (Spring, 1979), 37-75.

2. Robert Stein, "The Gold Fields of Alaska," *Review of Reviews,* XIII (June, 1896), 697-99.

3. The Klondike-Yukon literature is enormous. The best popular, accurate work is by Pierre Burton, *Klondike: Life and Death of the Last Great Gold Rush* (Toronto: McClelland & Stewart, 1958). One of the better contemporary works is Miner Bruce, *Alaska* (New York: G. P. Putnam Sons, 1899). I have relied heavily for my general comments on the Dawson newspapers of 1899-1904; documents in the Yukon Archives, Whitehorse; the rich Alaska and Yukon materials in the house of the Sisters of St. Ann, Victoria; and the many contemporary guidebooks and pamphlets.

4. William Ogilvie, *Early Days on the Yukon* (London: John Lane, 1913), pp. 118-22. The Ogilvie book is the historian's dream; a well-written, accurate account by a participant-observer whose integrity has never been questioned. Ogilvie was Dominion Land Surveyor and the Canadian government's leading authority in the region on land issues.

5. I have long been interested in Ladue's career. He was born in Plattsburgh, New York, of French Canadian descent, was in Alaska at least by 1882, and became a leading figure in Dawson's early history. Later he wrote two books on Klondike travel and prospecting. A good contemporary account of his activity with photos of him and his Dawson sawmill is in the *Klondike News,* July 1, 1898.

6. See, for example, issues of July 30 (impact in Phoenix and Prescott), July 31 (madness at Albuquerque), and August 6 (report from Chief Clerk of the San Francisco Mint).

7. *Tombstone Prospector,* November 19, 1897; *Florence Tribune,* November 20, 1897, both newspapers copying the original story from the *Arizona Daily Star.*

8. The *Jerome Mining News* of January 15, 1898, reported that Nellie had left Yuma, bound for the Klondike, via San Francisco. For Finley, see *Coconino Sun,* November 7, 1903; for Cunningham, *Bisbee Daily Review,* April 1, 1948. Yuma *Arizona Sentinel* of March 12, 1898, reported that Nellie had left Victoria the previous week, headed for the Klondike country.

9. Much of the female impact on the Klondike is well summarized in Melanie J. Mayer, *Klondike Women: True Tales of the 1897-98 Gold Rush* (Columbus: Ohio University Press, 1989). An early history and photos of Belinda Mulrooney are in *Klondike News,* July 1, 1898. "Firsts" were every-

where, and reports were usually erroneous. The *Los Angeles Herald* of August 2, 1897, announced that Mrs. Bessie Thomas left San Franciso, intending to be the first restaurant woman on the Klondike; at least four other women were already in business there.

10. This 1½ column interview was in the issue of February 15, 1898, p. 8.

11. A copy of the letter, dated February 22, 1898, is in the Cashman Papers. The letter was also printed in the *Victoria Daily Times* of January 10, 1925, shortly after Nellie's death.

12. *Daily British Colonist*, March 9, 1898.

13. Her travel plans, and changes, are in the Yuma *Arizona Sentinel*, April 16, 1898.

14. Henry F. Woods, *God's Loaded Dice. Alaska, 1897-1930* (Caldwell: Caxton Press, 1948), pp. 64, 132-33.

15. *Ibid.*

16. Some of these details appear in an article shortly after her death in the *British Daily Colonist*, January 11, 1925; the Yuma *Arizona Sentinel* of April 16, 1898, published a letter that she had written from Lake Bennett to a friend in Bisbee.

17. The quote is from Tyrrell, "Concentration of Gold in the Klondike," *Economic Geology*, II (June, 1907), 343-49. Tyrrell had earlier done major field work, and his first important writing was "Gold-Bearing Alluvial Deposits of the Klondike District," *Transactions, Institution of Mining & Metallurgy* [London], VIII (1899-1900), 217-29. The handiest set of sound sources on the geology and early mining are in *Yukon Territory*, Memoir 284, Geological Survey of Canada (Ottawa, 1957). This contains reprints of field reports first published from 1898 through 1933.

18. *Yukon Territory, City of Dawson* [Illustrated] (Dawson: American Journal of Industry, 1901), is a 73-page pamphlet with excellent photos, names of thousands of mine claimants, lists of merchant houses and newspapers, hotels, and churches.

19. *An American Missionary: A Record of the Work of Rev. William H. Judge, S.J.* (Ossining, N.Y.: Catholic Foreign Mission Soc., 1907); Sister Marie de L'Ange-Gardien, *En Alaska: L'Oeuvre des Soeurs de Sainte-Anne* (Montreal: Arbour & Laperle, 1900), pp. 11-15.

20. Dr. W. T. Barrett arrived in Dawson in early June of that year to become head physician at St. Mary's Hospital; he was to know Nellie for years, and he was her physician at the time of her death in Victoria in 1925. See Barrett, "Reminiscences of Early Klondyke Days," *St. Ann's Journal*, April, 1947, pp. 9-13.

21. Clum, *Arizona Historical Review*, pp. 27-28.

22. *Ibid.; Yukon Midnight Sun*, October 28, 1898, Cassiar Restaurant; *Anchorage Daily News*, January 19, 1925, Delmonico Restaurant.

23. News clipping, dated September, 1898, Sharlot Hall Museum, Prescott, entitled "Heard from in Klondike."

24. "Further Report of William Ogilvie," *Sessional Papers* (No. 87B.-87C.), Ottawa: Parliament, 1899.

25. File No. 187, Yukon Archives, Whitehorse.

26. Senkler to Secretary, Dept. of the Interior, Ottawa, July 8, 1901, copy in Yukon Archives.

27. Summarized in *Daily Klondike Nugget*, April 30, 1900; see also *Dawson Daily News, April 30, 1900.*

28. In a wishy-washy memo of October 17, 1900, to the Secretary of the Interior in Ottawa, Senkler claimed that granting compensation would set a bad policy for the government; Yukon Archives.

29. Barwell statement and legal correspondence concerning the compromise are in Yukon Archives.

30. Oath dated July 8, 1901, Gold Commissioner's Office, in Yukon Archives.

31. Memorandum of E. Senkler, Gold Commissioner, November 30, 1901, Yukon Archives.

32. News clipping in the Cashman Papers.

33. Lockley interview data published shortly after Nellie's death in *Anchorage Daily News*, January 19, 1925. See also Clum, *Arizona Historical Review*, pp. 27-30; Mayer, *Klondike Women*, p. 57.

34. *Alaska-Yukon Directory and Gazetteer, 1902* (Seattle: Alaska Yukon Dir. Co., 1902), pp. 242, 256. One never knows when directory data was collected, but this 1902 publication most likely represents information of early to mid-1901.

35. *Alaska-Yukon Directory, 1902*, pp. 175-76.

36. Sister Margaret Cantwell, *North to Share: The Sisters of Saint Ann in Alaska and the Yukon Territory* (Victoria: Sisters of St. Ann, 1992), pp. 86-92 is a good summary of Dawson's health and medicine. The disgusting garbage is mentioned on p. 1, *Dawson Daily News*, August 31, 1899.

37. Records of St. Mary's Hospital, Dawson, in Archives, Sisters of St. Ann, Victoria. Good wishes were extended in the *Dawson Daily News*, December 27, 1902.

38. Sister Mary Theodore, "The Sisters of St. Ann on North Pacific Shores," typescript, Part 2, pp. 420-21; Archives, Sisters of St. Ann, Victoria.

39. The account has been reprinted in a dozen or so Alaskan publications; see, for example, *Alaska Weekly*, February 13, 1925, and *Alaska Life*, April, 1945, pp. 51-52.

40. News notices about her Arizona visit are in the *Arizona Daily Citizen*, October 21, 1903, and *Coconino Sun*, November 7, 1903. Several documents and letters in the Cashman Papers deal with her relationship with the Cunningham children.

41. Some details are in *Dawson Daily News* of January 26, 1904, and *Whitehorse Daily Evening Star*, April 4, 1904.

42. *Yukon Territory: Its History & Resources* (Ottawa: Department of the Interior, 1916), Chapter V, "Placer Mining."

43. The *Yukon Catholic* of October, 1902, carried an advertisement for the hospital with this claim, also boasting of "best heated" and "best venti-

lated." Of the six physicians, Dr. W. T. Barrett was in charge, and heading the Sisters of St. Ann there was Sister Superior Zenon.

44. From an article by Bernice Cosulich several days after Nellie's death; *Arizona Daily Star,* January 11, 1925. Miss Cosulich had interviewed Nellie several months earlier during an Arizona stay.

45. An outstanding work on the history of central Alaska is Melody Webb's *The Last Frontier* (Albuquerque: University of New Mexico Press, 1985). The work is a combination of solid scholarship, clear writing, and reasonable interpretation. See also William. R. Cashen, *A Brief History of Fairbanks* (Fairbanks: n. p., 1971, especially pp. 2-6.

46. *Tanana Valley, Alaska* (Fairbanks: Chamber of Commerce, 1907), a 40-page pamphlet, copy in Huntington Library. Two useful articles on Fairbanks in 1906 appear in Herbert L. Heller, *Sourdough Sagas* (Cleveland: World Publication Co., 1967).

47. The *Dawson Daily News* of July 1, 1904, listed Nellie as operator of a Fairbanks restaurant. Nellie's statement of her work and profits are in the *Anchorage Daily News,* January 19, 1925.

48. Clara Heintz Burke, *Doctor Hap* (New York: Coward-McCann, Inc., 1961), pp. 91-92. Clara met Nellie in 1907; the above account was through Nellie's very good friend, Deaconess Carter.

49. Mission Diary in Archives, Sisters of St. Ann, Victoria.

Chapter Seven
Fulfillment on the Koyukuk

1. Webb, *Last Frontier,* pp. 139-41.

2. Josiah E. Spurr, "Geology of the Yukon Gold District," in *18th Annual Report, United States Geological Survey* (Washington: GPO, 1898), p. 125; F. C. Schrader, "Preliminary Report on a Reconnaissance Along the Chandlar and Koyukuk Rivers, Alaska, in 1899," *21st Annual Report, United States Geological Survey* (Washington: GPO, 1900, Part II, esp. pp. 447-63.

3. This appeared in the London publication *Wide World Magazine,* V (September, 1900), 608-16.

4. *Nome Gold Digger,* October 10, 1900; *Arctic Weekly Sun* (Nome), July 29, 1900, Bettles story.

5. There is good travel and climate detail on the Koyukuk in F. C. Schrader, "Recent Work of the U.S. Geological Survey in Alaska," *Bulletin of the American Geographical Society,* XXXIV (No. 1, 1900), 1-16.

6. Hudson Stuck, *Voyages on the Yukon and Its Tributaries* (New York: C. Scribner's Sons, 1917), quote from p. 314.

7. A comprehensive look at the geology of the Yukon-Koyukuk Basin is in a special issue of *Journal of Geophysical Research,* XCIV (November 10, 1989).

8. This is the point of view in John T. Dillon, "Source of Lode-and Placer-Gold Deposits of the Chandalar and Upper Koyukuk Districts, Alaska,"

Alaska Department of Natural Resources, Open-File Report 158 (September, 1982).

9. John C. Antweiler, et al, "Studies of Gold in the Chandalar and Koyukuk Districts, Wiseman and Bettles Quadrangles," *U.S. Geological Survey Circular 945* (1985).

10. The Cashman Papers contain many documents regarding Nellie's claims around Nolan Creek. The lease with Richmond presents a chronological problem. It was dated July 27; yet on July 8 she arrived at Nulato to visit the Sisters of St. Ann, on her way north for the winter. It is possible that the lease was signed in Fairbanks on June 27, enabling her to get from Fairbanks to Nulato by July 8.

11. Undated statement by McCord, probably before 1919, when he was writing a document to assist her in getting financing; Cashman Papers. Not many prospectors had luck in the Chandalar District, and all newcomers were frozen out. The best grounds "were in the hands of a few men who are holding it for speculation"; Chena *Tanana Miner,* September 15, 1907.

12. The article was reprinted in the *Whitehorse Weekly Star* of October 12, 1906.

13. Stuck, *Voyages,* p. 355.

14. Burke, *Doctor Hap,* pp. 90-92.

15. *Fairbanks Daily News-Miner,* July 21, 1908.

16. *Fairbanks Daily Times,* July 22, 1908.

17. Issue of November 21, 1908, contains details of her Arizona trip as well as her overland travel plans to Nolan Creek.

18. George M. Hill, "The Koyukuk: One of the Richest Districts in the Far North," *Alaska-Yukon Magazine,* VIII (June, 1909), 210-13.

19. *Fairbanks Weekly Times,* August 7, 1909.

20. It is difficult to distinguish these names in the data published in placename digests or works on local history. Even the mining claims contain confusion about the places.

21. For Cunningham around Nolan Creek, information from "Location Notices," a detailed list of all Nellie's Nolan Creek properties, Cashman Papers. The Cochise County quit claim deed was signed on December 7, 1909; Cochise County, Deed of Mines, Book 25, 460.

22. McCord statement, undated but probably 1919, in Cashman Papers.

23. *Mineral Resources of the United States, 1914: Metals* (Washington: GPO, 1916), Koyukuk District on p. 135.

24. *Mining & Scientific Press,* September 12, 1914.

25. A. G. Maddern, *The Koyukuk-Chandalar Region* (Washington: GPO, 1913), Nolan Creek data on pp. 92-94. This publication was Bulletin 532 of the U.S. Geological Survey.

26. *Engineering & Mining Journal,* June 12, 1915.

27. The year 1916 is well covered in the Seattle *Alaska and Northwest Mining Journal,* in particular the issue of March, p. 59. The issue of November, p. 132, contains the production chart.

28. *Whitehorse Star,* October 6, 1916, trip details.

29. The Seattle interview, datelined October 29, appeared in the *Alaska Daily Empire*, November 2, 1916.

30. Stuck, *Voyages on the Yukon*, p. 358.

31. *Tanana Miner*, April 5, 1908.

32. Emil Hurja comment in *Mining & Scientific Press*, September 12, 1914.

33. Robert Marshall, *Arctic Village* (New York: Literary Guild, 1933).

34. Most of these comments are based on Marshall's fascinating study. Another useful work is Margaret E. Murie, *Two in the Far North* (New York: Alfred A. Knopf, 1962), which is good for the geography of the area, but does not examine the people in much depth.

35. Burke, *Doctor Hap*, p. 213.

Chapter Eight
End of the Paystreak

1. *Cordova Daily Times*, January 4, 1921.

2. These were all from one interview; *Fairbanks Daily Times*, August 22, 1908.

3. *Daily Alaskan*, June 6, 1919.

4. *Alaska Daily Empire*, December 22, 1920.

5. *Cordova Daily Times*, January 4, 1921.

6. *Alaska Daily Empire*, January 14, 1921.

7. Copies of these and other testimonials are in the Cashman Papers.

8. The plans to incorporate are in *Pathfinder*, October, 1922, p.10. The stock certificate is in the Rasmuson Library, University of Alaska-Fairbanks.

9. *Sunset*, XLVI (May, 1921), 48, in an article entitled "Interesting Westerners."

10. *Pathfinder*, January, 1924, p. 14.

11. Letter dated April 15, 1923, in Cashman Papers.

12. *Arizona Daily Star*, January 18, 1924.

13. Rev. Fr. Thomas A. Doyle to Sister Mary Loyola, July 7, 1971, in Cashman Papers. Fr. Doyle, writing from Tombstone, does not give the date of this donation, writing only "1923," so it could be from the winter of 1922-23 or winter of 1923-24.

14. *Alaska Weekly*, September 5, 1924.

15. Chronicles of St. Peter Claver Mission, entries for August 2-5, in Archives, Sisters of St. Ann, Victoria. Some details of the last trip are also in *Stroller's Weekly* (Juneau), August 30, 1924.

16. *Alaska Weekly*, October 3, 1924.

17. Personal checks, her will, and reminiscences of several of the sisters in attendance in her last days are in the Cashman Papers.

18. The autopsy information is from a statement by Sister Mary Patrick, who was on the nursing staff at the time; Cashman Papers. Death notice in Victoria *Daily Colonist*, January 6, 1925, with plans for the funeral ceremonies. The death certificate, filed at the Ministry of Health, British Columbia, Victoria, gives her age as eighty.

19. Quote from *Pathfinder*, January, 1925, p. 16. Some of the obituaries are Victoria *Daily Colonist*, January 11, 1925; *New York Times*, January 8, 1925, p. 25; *San Francisco Chronicle*, January 7, 1925, p. 8; *Tombstone Epitaph*, January 9, 1925; *Arizona Daily Star*, January 7, 1925; *Anchorage Daily News*, January 19, 1925.

20. Issue of January 31, 1925, p. 226.

21. This point of view was expressed by Sister Mary Loyola in a letter to J. G. Magna of the Alaska Miner's Association, June 2, 1972. She was following up on this question, but never settled it; letter in Cashman Papers.

22. This is in Part 2, p. 418, of Sister Mary Theodore's "The Sisters of St. Ann on North Pacific Shores," typescript in Archives, Sisters of St. Ann, Victoria.

23. Diary in Archives, Sisters of St. Ann, Victoria. The account was a bit confused, as Nellie died at St. Joseph's Hispital, Victoria.

24. Victoria *Daily Colonist*, January 11, 1925.

25. Philip S. Smith, *Mineral Resources of Alaska, U. S. Geological Survey, Bulletin 797* (1929), pp. 21-22.

BIBLIOGRAPHY

Of the materials used for this book, by far the richest and most numerous came from the Cashman Papers in the Archives of the Sisters of St. Ann in Victoria. The Archives not only contain important Alaskan and Canadian materials relating to mission work, but also many other materials that concern Nellie Cashman's years in the north country. Furthermore, in the 1970s Sister Mary Loyola made a concerted effort to obtain copies of most vital correspondence, articles, and reports from other depositories in the United States and Canada.

MANUSCRIPTS/PRIMARY SOURCES

Bisbee, Cochise County, Arizona. Mining, tax, real estate, census and probate records.

Cork, Ireland, Cork County Library. Reference files of the parish of Midleton.

Edmonton, Alberta, Provincial Archives. Hudson's Bay Co. records.

Fairbanks, Elmer E. Rasmuson Library, University of Alaska-Fairbanks. Nellie Cashman file.

Los Angeles, California, Natural History Museum. Maps, photos, census, real estate records.

San Marino, California, Huntington Library. Fred Dodge Papers; Stewart Lake Collection; various map, photo, and rare book collections.

Tucson, Arizona, Arizona Historical Society. Misc. Cashman material; John Pleasant Gray Reminiscences; various biographical files.

Victoria, British Columbia, Archives of the Sisters of St. Ann.

Nellie Cashman Papers; various collections regarding British Columbia, Yukon Territory, and Alaska.
Victoria, British Columbia, Provincial Archives. Mining claims, tax records, census files.
Whitehorse, Yukon Territory, Provincial Archives. Mining claims; legal controversies.

GOVERNMENT PUBLICATIONS

Antweiler, John C. et al. *Studies of Gold in the Chandalar and Koyukuk Districts, Wiseman and Bettles Quadrangles.* United States Geological Survey Circular 945, 1985.
Dawson, George M. *Report on an Exploration in the Yukon District, N.W.T., and Adjacent Northern Portion of British Columbia, 1887.* Montreal: Dawson Bros., 1888, by Authority of Parliament.
Dillon, John T. *Source of Lode- and Placer-Gold Deposits of the Chandalar and Upper Koyukuk Districts, Alaska.* Alaska Department of Natural Resources, Open-File Report 158, September, 1982.
"Further Report of William Ogilvie." *Sessional Papers,* No. 87B.-87C., Ottawa: Parliament, 1899.
Maddern, A. G. *The Koyukuk-Chandalar Region.* Bulletin 532, U. S. Geological Survey, 1913.
Mineral Resources of the United States. Washington: U. S. Geological Survey, 1914.
Schrader, F. C. "Preliminary Report on a Reconnaissance Along the Chandlar and Koyukuk Rivers, Alaska, in 1899." In *21st Annual Report, United States Geological Survey.* Washington: Government Printing Office, 1900, Part II.
Smith, Philip S. *Mineral Resources of Alaska. . .1926.* Bulletin 797, U. S. Geological Survey, 1929.
Spurr, Josiah E. "Geology of the Yukon Gold District." In *18th Annual Report, United States Geological Survey.* Washington: Government Printing Office, 1898.
Yukon Territory. Memoir 284, Geological Survey of Canada, 1957.
Yukon Territory: Its History & Resources. Ottawa: Department of the Interior, 1916.

BOOKS

Alaska-Yukon Directory and Gazetteer, 1902. Seattle: Alaska-Yukon Directory Co., 1902.

An American Missionary: A Record of the Work of Rev. William H. Judge, S.J. Ossining, N.Y.: Catholic Foreign Mission Society, 1907.

Andrews, Clarence L. *Wrangell and The Gold of the Cassiar*. Seattle: Luke Tinker, 1937.

Bancroft, H. H. *History of Arizona and New Mexico*. San Francisco: History Company, 1889.

Bancroft, H. H. *History of British Columbia, 1792-1887*. San Francisco: History Company, 1887.

Bancroft, H. H. *History of Nevada, Colorado, and Wyoming*. San Francisco: History Company, 1890.

Brophy, Frank Cullen. *Arizona Sketch Book*. Phoenix: Ampco Press, 1952.

Burchell, R. A. *The San Francisco Irish, 1848-1880*. Berkeley: University of California Press, 1980.

Burke, Clara Heintz. *Doctor Hap*. New York: Coward-McCann, Inc., 1961.

Burton, Pierre. *Klondike: Life and Death of the Last Great Gold Rush*. Toronto: McClelland & Stewart, 1958.

Cantwell, Sister Margaret. *North to Share: The Sisters of Saint Ann in Alaska and Yukon Territory*. Victoria: Sisters of St. Ann, 1992.

Cashen, William R. *A Brief History of Fairbanks*. Fairbanks: n.p., 1971.

Chaput, Donald. *Virgil Earp: Western Peace Officer*. Encampment, Wyoming: Affiliated Writers of America, 1994.

Chaput, Donald et al. *The Mines of Baja California Norte*. Los Angeles: Natural History Museum, 1992.

Chase, Will H. *The Trail Blazers of Bygone Days*. Kansas City: Burton Publishing Co., 1951.

Heller, Herbert L. *Sourdough Sagas*. Cleveland: World Publishing Co., 1967.

History of Arizona: Biographical. Phoenix: Record Publishing, 1930.

Hulse, James W. *Lincoln County, Nevada, 1864-1909*. Reno: University of Nevada Press, 1971.

L'Ange-Gardien, Sister Marie. *En Alaska: L'Oeuvre des Soeurs de Sainte-Anne*. Montreal: Arbour & Laperle, 1900.

Ledbetter, Suzann. *Nellie Cashman: Prospector and Trailblazer*. El Paso: Texas Western Press, 1983.

MacLysaght, Edward. *More Irish Families*. New York: Barnes & Noble, Inc., 1960.

McKenna, James A. *Black Range Tales*. New York: Wilson-Erickson, Inc., 1936.

McKenney's Business Directory . . . Central and Southern California, Arizona, New Mexico, Southern Colorado and Kansas. San Francisco: Pacific Press, 1882.

Mayer, Melanie J. *Klondike Women: True Tales of the 1897-98 Gold Rush*. Columbus: Ohio University Press, 1989.

Miller, Kerby A. *Emigrants and Exiles: Ireland and the Irish Exodus to North America*. New York: Oxford University Press, 1985.

Murie, Margaret E. *Two in the Far North*. New York: Alfred A. Knopf, 1962.

Marshall, Robert. *Arctic Village*. New York: Literary Guild, 1933.

Nolan, Janet. *Ourselves Alone: Women's Emigration from Ireland, 1885-1920*. Lexington: University Press of Kentucky, 1989.

Ogilvie, William. *Early Days on the Yukon*. London: John Lane, 1913.

Patterson, R. M. *Trail to the Interior*. New York: Wm. Morrow & Co., 1966.

Sherman, James E. & Barbara. *Ghost Towns and Mining Camps of New Mexico*. Norman: University of Oklahoma Press, 1974.

Sonnichsen, C. L. *Tucson: Life and Times of an American City*. Norman: University of Oklahoma Press, 1982.

Stuck, Hudson. *Voyages on the Yukon and Its Tributaries*. New York: C. Scribner's Sons, 1917.

Tanana Valley, Alaska. Fairbanks: Chamber of Commerce, 1907.

Webb, Melody. *The Last Frontier*. Albuquerque: University of New Mexico Press, 1985.

Weber, Francis. *The Peninsular California Missions, 1808-1880*. Los Angeles: The Author, 1979.

Williams, R. Hal. *The Democratic Party and California Politics, 1880-1896*. Stanford: Stanford University Press, 1973.

Woods, Henry F. *God's Loaded Dice. Alaska, 1897-1930.* Caldwell: Caxton Press, 1948.

Yukon Territory, City of Dawson [Illustrated]. Dawson: American Journal of Industry, 1901.

ARTICLES

Auer, Louise Cheney. "Nellie Cashman and Her Lost Gold." *Desert Magazine,* XXVI (October, 1963), 24-26.

Barrett, W. T. "Reminiscences of Early Klondyke Days." *St. Ann's Journal,* April, 1947, 9-13.

Blake, W. P. "The Geology and Veins of Tombstone, Arizona." *Transactions, American Institute of Mining Engineers,* X (1882), 334-45.

Clum, John P. "Nellie Cashman." *Arizona Historical Review,* III (January, 1931), 9-34.

Ellis, George M. "Sheriff Jerome L. Ward and the Bisbee Massacre of 1883." *Journal of Arizona History,* XXXV (Autumn, 1994), 315-42.

Hogan, Charles. "Did Nellie Cashman Find Gold?" *The West,* May, 1965, p. 17.

Hudson, J. "Our Klondike Failure." *Wide World Magazine,* V (September, 1900), 608-16.

"Interesting Westerners." *Sunset,* XXLVI (May, 1921), 48.

Mercier, Laurie K. "We are Women Irish." *Montana Magazine of History,* XLIV (Winter, 1994), 29-41.

Pepper, Choral. "An Angel and the Valley Gold." *Desert Magazine,* XXXVIII (December, 1975), 24-27.

Schrader, F. C. "Recent Work of the U. S. Geological Survey in Alaska." *Bulletin of the American Geographical Society,* XXXIV (No. 1, 1900), 1-16.

Stein, Robert. "The Gold Fields of Alaska." *Review of Reviews,* XIII (June, 1896), 697-99.

Tyrrell, J. B. "Concentration of Gold in the Klondike." *Economic Geology,* II (June, 1907), 343-49.

Tyrrell, J. B. "Gold-Bearing Alluvial Deposits of the Klondike District." *Transactions, Institution of Mining & Metallurgy* (London), VIII (1899-1900), 217-29.

Underhill, Lonnie E. "The Tombstone Discovery: Recollections of Ed Schieffelin & Richard Gird." *Arizona and the West,* XXI (Spring, 1979), 37-75.

PERIODICALS/PROFESSIONAL JOURNALS

Alaska and Northwest Mining Journal (Seattle)
Alaska-Yukon Magazine (Seattle)
Arizona Quarterly Illustrated (Phoenix)
Engineering & Mining Journal (New York City)
Mining & Scientific Press (San Francisco)
Journal of Geophysical Research (Washington, D.C.)
Pathfinder (Valdez, Alaska)

NEWSPAPERS

Alaska Daily Empire (Juneau)
Alaska Weekly (Seattle)
Anchorage Daily News
Arctic Weekly Sun (Nome)
Arizona Daily Star (Tucson)
Arizona Miner (Prescott)
Arizona Sentinel (Yuma)
Bisbee Daily Review
Cariboo Sentinel (Barkerville, Br. Col.)
Chicago Tribune
Cordova Daily Times
Daily Alaskan (Skagway)
Daily Arizona Citizen (Tucson)
Daily British Colonist (Victoria)
Dawson Daily News
Ely Record (Pioche)
Fairbanks Daily News-Miner
Fairbanks Times
Globe Silver-Belt
Klondike News (Dawson)
Klondike Nugget (Dawson)
Los Angeles Herald
Los Angeles Times
Lower Californian (Ensenada, B.C.)

New York Times
Nome Gold Digger
Phoenix Daily Herald
Pioche Daily Record
San Diego Sun
San Diego Union
San Francisco Chronicle
Stroller's Weekly (Juneau)
Tanana Miner (Chena)
Tombstone Epitaph
Tombstone Nugget
Tombstone Prospector
Whitehorse Star
Yukon Catholic (Dawson)
Yukon Midnight Sun (Dawson)

INDEX

Alaska, mining, 116-19, 120-56; Ft. Wrangell, 21-22
Alcohol, in Koyukuk Basin, 139-40
Allakaket, Episcopal Mission, 122, 126
Alsek, mining, 111, 113
American Hotel, Tombstone, 51-53
Arcade Restaurant, Tombstone, 41, 45
Archibald, Miles, 87, 112
Arizona, mining, milling, 127-74
Arizona Daily Star, 35, 69, 80

Barrett, Dr. W. T., Yukon, 101, 150-52
Barwell, C. S. W., surveyor, 106
Belcourt & McDougal, attorneys, 106
Benson, Ariz., railroad, 45
Bettles, trading post, 122, 126
Bettles, Gordon, 121-22
Bisbee, Ariz., 46, 65-67, 128
Bisbee Hotel, 46
Bisbee Massacre, 65-67
Black Range Mts., New Mex., 76-77
Bonanza Creek, Yukon, 90, 103, 105, 106-09
Boston, Cashman family, 5
British Columbia, mining, 16-26; *see also* Victoria
Brophy, Frank C., AZ, 70
Brophy, William H., AZ, 69-70
Buffalo Hotel, Globe, 86
Bullionville, Nev., 7, 11
Burke, Dr. Hap, Alaska, 139-40

Caddy, W. A., Tombstone, 49
Carter, Deaconess Clara, 117-18, 126-27
Cashman, Mrs. Frances (Fanny), 2-8, 12-13, 26, 47
Cashman, Nellie [Ellen], mining claims, 92, 64-65, 70-71, 84, 103-15, 124-25, 131-32; in Arizona, 27-74; Baja California, 56-63; British Columbia, 16-26; Cripple Creek, 86; Fairbanks, 116-19; Globe, 86; Harqua Hala, 77-83; Ireland, 1-7; Jerome, AZ, 85; Juneau, 85, 142-43; Kingston, New Mex., 75-77; Koyukuk, Alaska, 120-56; Minas Prietas, 85-86; Montana, 84; Nevada, 7-15; Nogales, AZ, 71-72; Prescott, 84; San Francisco, 5-7; Tombstone, 34-75; Tucson, 29-34, 73, 82-83, 193; Victoria, B. C., 94-95, 150-52; Yuma, 86-88; Yukon, 89-115
Cashman, Patrick, 2-3
Cashman House, Kingston, 76
Cassiar District, B.C., mining, 16-26
Cassiar Restaurant, Dawson, 102
Catholics; *see* Roman Catholics
Chandlar (Chandalar), Alaska, mining, 123, 125
Charleston, AZ, 42
Chena River, Alaska, 116-18
Chilkoot Pass, 96-98
Choquette, Alex. "Buck," Canadian prospector, 18
Christy, Mary, Bisbee, 47
Churches; *see* Episcopal; Roman Catholics
Claude, Sister Mary, Victoria, 150
Clum, John, Arizona, Alaska, 29, 33-34, 38-39, 68, 102
Cochise County, 38, 44, 46, 64
Cochise Mining District, 64
Colorado River, transportation, 86-88
Contention, AZ, 42, 60
Copper mining, Bisbee, 46-67; Globe, 86; Jerome, 85
Coyle, James, 82
Crime; *see* Violence

Cripple Creek, Colo., 86
Cunningham, Mrs. T. J. (Fanny), 1-7, 47, 50-52, 55-56, 70
Cunningham, Mamie, 87
Cunningham, Mike, 55, 69-70, 84, 92, 112, 128, 132, 146, 152-53
Cunningham, Tom (Sr.), 6, 34, 47
Cunningham, Tom (Jr.), 73, 94-96, 103, 109

Dawson, Yukon Terr., 89-115
Dawson Daily News, 111
Davie, Justice Theodore, 94
Daily Arizona Citizen, 33
Daily Arizona Star, 28-29
Daily British Colonist, 18, 23, 25, 94
Dease Lake, Br. Col., 19, 22-26
Delmonico Restaurant, Dawson, 102; Nogales, 71-72; Tuscon, 29-30, 34
Delmonico Lodging House, Tombstone, 47, 49
Diamond mining, South Africa, 83-84
Dodge, Fred, Arizona, 68
Doheney, Ed, New Mexico, 76-77
Donovan Hotel, Dawson, 103
Dragoon Mts., AZ, 41, 49
Dubin, Sam, Alaska, 149

Earp bros., Tombstone, 33, 37, 50
Earp, Virgil, police chief, 37, 50
Earp, Wyatt, mining, 78, 83
Elite Restaurant, Prescott, 85
Episcopal Church, Allakaket, 122, 126; Fairbanks, 117

Fairbanks, hospital, 117-18, 150; mining, 116-19; mining equip., 127-28, 142, 149
Fairbanks Daily Times, 129
Farthing, Annie Cragg, Alaska, 117
Finley, James, Tucson, 92
Fort Wrangell, Alaska, 21-25

Gage, E. B., mining supt., 68-69, 71
Geology, Cassiar Dist., 22-23; Harqua Hala, 80-82; Koyukuk, 123-24; Nolan Creek, 131, 133-34; Tombstone, 31; Yukon, 53-54
"Gisna," steamer, 60
Gold mining, Baja California, 56-63; British Columbia, 16-26; Fairbanks, 116-19; Harqua Hala, 77-83; Koyukuk, 120-56; Minas Prietas, 85-86; Yukon, 89-115
Goodfellow, Dr. George, AZ, 65-66, 86
Gray, John Pleasant, AZ, 51
Guaymas, Mexico, 56-63

Hammond Creek, mining, 133
Handy, Dr. J. C., Tucson, 63
Harqua Hala, gold mining rush, 77-83
Hart, John, mining diary, 61
Heintz, Clara, Alaska, 126-27
Heith, John, lynched, 65-66
Hensley, James, AZ, 64
Hospitals, Dawson, 101-102, 115; Fairbanks, 117-18; Tombstone, 41-44, 46; Tucson, 30; Victoria, 26, 150-55
Hotel Cashman, Yuma, 87-88
Howard, Frank E., Alaska, 127
Huachuca Mts., AZ, 42, 64
Hudson, J., Alaska, 121
Hulse, James, quoted, 8
Hurja, Emil, mining writer, 133, 136

Ireland, Cashman family, 1-7
Irishmen, Dawson, 115-16; Nevada, 10-11; New Mexico, 76-77; Tombstone, 38-40, 44, 49-50
Indians, British Columbia, 18; Koyukuk, 122, 130, 137; Yaqui, 62

Jacobs & Sultan, Pioche, 14
Jerome, AZ, 85
Joyce, Milt, Baja Calif., 60; Oriental Saloon, 52
Judge, Fr. William, Dawson, 101
Juneau, Alaska, 85, 142-43

Kavanaugh, Thomas, AZ, 64
Kingston, New Mex., silver mining, 75-77
Klondike, Yukon Terr., 89-115
Korka, I. A., Tombstone, 45
Koyukuk Basin, Alaska, mining, 120-56
Koyukuk River, desc., 122

Ladue, Joseph, Yukon, 90-91
Laketon, Br.Columbia, 22
Lawing, Nellie, Alaska, 144
Lazard, N. B., Tombstone, 68
Leach, C. W., mining official, 68-69

Leahy, David, Nevada, 8
Lincoln County, Nev., mining, 7-15
Lockley, Fred, quoted, 20, 107, 156
Los Angeles, 29, 40

McCarty, Dennis, AZ, 64, 75
McCord, J. W., Alaska, 132
McCullough, ___, prospector, 18-19
McIntyre, R. E., Koyukuk Basin, 131,32, 140, 152
McQuade, Malcolm, AZ, 64, 75
Maddren, A. G., geologist, 133
Maish, Fred, Tucson, 73
Mark, Sister Mary, Victoria, 150-52
Marshall, Robert, quoted, 136-39
Miner's Dream, AZ, 75
Midnight Sun Mining Co., Alaska, 145, 153
Mining, Alaska, 116-56; Baja California, 56-63; British Columbia, 16-26; Fairbanks, 116-19; Globe, 86; Harqua Hala, 77-83; Koyukuk Dist., 120-56; Minas Prietas, 85-86; Nevada, 7-15; New Mexico, 75-77; Tombstone, 30-75; Yukon, 89-115; Yuma, 86-88
Montana, 84
Mulege, Baja California, 56-63
Mulrooney, Belinda, Dawson, 103-05

New York City, 146
New York Times, 113
Nogales, AZ, mining, 72-73
Nolan, Alaska, 131
Nolan, John, Alaska, 124
Nolan Creek, geology, mining, 124-56
Northwest Mounted Police, 96, 109
Norwood, ___, mining inspector, Dawson, 103-05
Nulato, Alaska, Roman Catholic mission, 117, 119, 149-50

O'Boyle, Neal, New Mexico, 76
Ogilvie, William, Yukon Terr., 103-05
O'Hara, Kate, Tombstone, 34-35, 40
Oleson bros., mining, Alaska, 129

Palace Hotel, Tucson, 73
Panaca Flat, Nevada, 7-15
Parnell Mine, AZ, 64
Pascholy, Joseph, Tombstone, 41, 49, 65
Pedro, Felix, Fairbanks, 116
Phoenix, AZ, 78-83, 85
Phoenix *Daily Herald*, 78-79
Pima County, AZ, 31
Pioche, Nev., silver mining, 7-15, 19
Pioche Daily Record, 8, 10, 14, 19
Population, Baja California mines, 57; Dawson, 93-94, 100; Fairbanks, 117; Koyukuk, 121, 123, 138; Tombstone, 38
Prescott, mining, 84
Prostitution, in Koyukuk Basin, 137-38

Quinn, John, Harqua Hala mining, 81-83

Railroads, Arizona, 27-29, 33, 45; Fairbanks, 117; Mexico, 58-59; Nogales, 72-73
Reilly, Justice James, 44
Richmond, Volney, Fairbanks, 124
Riveroll, Manuel, Baja California, 58
Roman Catholics, Alaska, 118-19, 149-50; Arizona, 55-56, 149; British Columbia, 153-57; Dawson, 101-02, 110-11, 115; Nevada, 11; New Mexico, 77; San Diego, 55-56; Tombstone, 41-44, 49-50; Tucson, 29-30; Victoria, 26, 150-52
Rowe, ___, Episcopal Bishop, 135
Russ House, Tombstone, 49-50, 65, 70-71
Russell, M. E., Yukon Terr., 106

"S.S. Californian," 23
"S.S. Centennial," 95-96
"S.S. Otter," 26
Sacred Heart Church, Tombstone, 44
Saint Joseph's Hospital, Fairbanks, 150
Saint Joseph's Hospital, Victoria, 26, 150-52
Saint Mary's Hospital, Dawson, 101-02, 110
Saint Matthew's Hospital, Fairbanks, 118
San Diego, Roman Catholics, 55-56
San Diego Sun, 67
San Diego Union, 63
San Francisco, Cashman-Cunningham links, 5-7,19-21, 40, 56
Schieffelin, Ed, Alaska, 89; Tombstone, 30-31
Schreiber, Joseph, Baja California, 57
Seattle, 135, 143, 150

Senkler, E. A., Yukon Gold Commissioner, 105-07
Shibell, Charles, Harqua Hala mining, 78, 83
Silver mining, Arizona, 27-74; Nevada, 7-15; New Mexico, 75-77
Sisters of St. Ann, Alaska, 119; Dawson, 101-02, 110-11, 150; Juneau, 142; Nulato, 149-50; Victoria, 26, 150-56
Sisters of Saint Joseph, Tucson, 49-50
Smith, Mrs. E. J., 83
Smith, John, AZ, 71
Smith, Mark, Baja California mining, 60
Soap Creek, Yukon claims, 106-07
Sonora, Mexico, mining, 29-30
South Africa, diamond mining, 83-84
Stikine River, British Columbia, 16-26, 94-96
Strauss, Charles, Tucson, 58
Stuck, Hudson, Episcopal missionary, 123, 135
Sunset, Cashman interview, 145
Swift, Jennie, Tombstone, 35, 40
Sullivan, Mike, Harqua Hala mining, 77-83

Tanana Miner, 127
Tanana River, Alaska, mining, 116-18
"Teddy H.," Alaska steamer, 149
Theodore, Sister Mary, 153
Thibert, Henri, Canadian prospector, 18-19, 22
Thompson, Wm., mining claims, 105-07
Tombstone, geology, mining, milling, 31-74
Tombstone Epitaph, 34, 46, 49, 51, 73
Tombstone Nugget, 33, 34, 42, 45, 50; destroyed by fire, 52
Tucson, Cashman business, 27-30, 34; 82-83; 149
Turquoise Mining Dist., AZ, 64, 132
Tyrrell, J. B., geologist, 99

United States Army, Tombstone labor troubles, 69

Victoria, British Columbia, 18, 20-21, 24, 26, 94-95, 150-52
Violence, Bisbee, 65-67; Dawson, 109-10; Harqua Hala, 78; Koyukuk Basin, 123, 138-40; mining strike, 68-70; Tombstone, 35-38; Nevada, 9-10

Ward, Sheriff Jerome, Tombstone, 66, 68
Ward, Kate, Tombstone, 70
Wickersham, Judge James, 143
Williams, O. R., Alaska, 124
Wiseman, Alaska, 131, 133-34, 147-49
Wiseman Creek, 127, 131
Women, on mining frontier, 23-25, 60-63, 80-82, 92-93, 127-28, 145-46, 156
World War I, mining production, 141
Wrangell, Alaska, 17, 19, 21-22, 95; *see also* Fort Wrangell

Yukon Territory, mining, 89-115
Yukon River, 89-90
Yuma, AZ, 86-88
Yuma *Arizona Sentinel*, 91-92

Design by Lynn R. Bailey
Typography by Western Lore Graphics, Inc.
Lithography by Commercial Printers Inc., Tucson
Edition Binding by Roswell Bookbinding, Phoenix